J. Paulo Davim (Ed.)
Wear of Composite Materials
Advanced Composites

Also of Interest

Series: Advanced Composites.
J. Paulo Davim (Ed.)
ISSN 2192-8983
Published titles in this series:
Vol. 8: Hierarchical Composite Materials (2018) Ed. by K. Kumar, J. P. Davim
Vol. 7: Green Composites (2017) Ed. by J. P. Davim
Vol. 6: Wood Composites (2017) Ed. by A. Alfredo, J. P. Davim
Vol. 5: Ceramic Matrix Composites (2016) Ed. by J. P. Davim
Vol. 4: Machinability of Fibre-Reinforced Plastics (2015) Ed. by J. P. Davim
Vol. 3: Metal Matrix Composites (2014) Ed. by J. P. Davim
Vol. 2: Biomedical Composites (2013) Ed. by J. P. Davim
Vol. 1: Nanocomposites (2013) Ed. by J. P. Davim, C. A. Charitidis

Shape Memory Polymers
Kalita, Hemjyoti, 2018
ISBN 978-3-11-056932-2, e-ISBN 978-3-11-057017-5

Materials Science
Vol. 1: Structure
Gengxiang Hu, Xun Cai, Yonghua Rong, 2018
ISBN 978-3-11-049512-6, e-ISBN 978-3-11-049534-8

Materials Science
Vol. 2: Phase Transformation and Properties
Gengxiang Hu, Xun Cai, Yonghua Rong, 2018
ISBN 978-3-11-049515-7, e-ISBN 978-3-11-049537-9

Wear of Composite Materials

Edited by
J. Paulo Davim

DE GRUYTER

Editor
Prof. Dr. J. Paulo Davim
University of Aveiro
Department of Mechanical Engineering
Campus Santiago
3810-193 Aveiro, Portugal
pdavim@ua.pt

ISBN 978-3-11-035289-4
e-ISBN (PDF) 978-3-11-035298-6
e-ISBN (EPUB) 978-3-11-038782-7
ISSN 2192-8983

Library of Congress Cataloging-in-Publication Data
Names: Davim, J. Paulo, editor.
Title: Wear of composite materials / edited by J. Paulo Davim.
Description: Berlin ; Boston : De Gruyter, 2018. | Series: Advanced composites ; volume 10
Identifiers: LCCN 2018021906| ISBN 9783110352894 (print) | ISBN 9783110352986 (pdf) | ISBN 9783110387827 (epub)
Subjects: LCSH: Mechanical wear. | Composite materials--Deterioration.
Classification: LCC TA418.4 .W4175 2018 | DDC 620.1/1892--dc23 LC record available at https://lccn.loc.gov/2018021906

Bibliographic information published by the Deutsche Nationalbibliothek
The Deutsche Nationalbibliothek lists this publication in the Deutsche Nationalbibliografie; detailed bibliographic data are available on the Internet at http://dnb.dnb.de.

Typesetting: Integra Software Services Pvt. Ltd.
Printing and binding: CPI books GmbH, Leck
Cover image: gettyimages/thinkstockphotos, Abalone Shell

www.degruyter.com

Preface

Generally, wear is defined as "progressive loss of material from the operating surface of a body occurring as a result of relative motion at the surface". Wear is an important parameter in tribology, usually defined as "the science and technology of interacting surfaces in relative motion", which involves research and application of principles of friction, wear, lubrication and design.

This book comprises six chapters, which aims at providing recent information on wear of composites and related matters. Chapter 1 provides information on tribology of carbon fabric-reinforced thermoplastic composites. Chapter 2 is dedicated to the effect of glass fibre reinforcement and the addition of MoS2 on the tribological behaviour of PA66 under dry sliding conditions (a study of distribution of pixel intensity on the counterface). Chapter 3 provides a review on dry sliding wear behaviour of metal–matrix composites. Chapter 4 describes tribology of aluminium–matrix composites. Chapter 5 provides information on silver-based self-lubricating composite for sliding electrical contact (material design, preparation and properties). Finally, Chapter 6 is dedicated to tribology of CrC–NiCr cermet coatings.

The book can be used as a research book for final-year undergraduate engineering course or as a matter on wear of composites at the postgraduate level. In addition, this book can serve as a suitable reference for academics, researchers, materials, mechanical and manufacturing engineers, professionals in composites and related industries.

The editor acknowledges De Gruyter for this opportunity and for their professional support. Finally, I would like to thank all the chapter authors for their availability for this work.

Aveiro, Portugal, July 2018 J. Paulo Davim

https://doi.org/10.1515/9783110352986-201

Contents

Biographical Sketche of editor

J. Paulo Davim received his Ph.D. degree in Mechanical Engineering in 1997, M.Sc. degree in Mechanical Engineering (materials and manufacturing processes) in 1991, Mechanical Engineering degree (5 years) in 1986, from the University of Porto (FEUP), the Aggregate title (Full Habilitation) from the University of Coimbra in 2005 and the D.Sc. from London Metropolitan University in 2013. He is Eur Ing by FEANI-Brussels and Senior Chartered Engineer by the Portuguese Institution of Engineers with an MBA and Specialist title in Engineering and Industrial Management. Currently, he is Professor at the Department of Mechanical Engineering of the University of Aveiro, Portugal. He has more than 30 years of teaching and research experience in Manufacturing, Materials, Mechanical and Industrial Engineering, with special emphasis in Machining & Tribology. He has also interest in Management, Engineering Education and Higher Education for Sustainability. He has guided large numbers of post-doc, Ph.D. and master's students as well as coordinated & participated in several financed research projects. He has received several scientific awards. He has worked as evaluator of projects for international research agencies as well as examiner of Ph.D. thesis for many universities. He is the Editor in Chief of several international journals, Guest Editor of journals, books Editor, book Series Editor and Scientific Advisory for many international journals and conferences. Presently, he is an Editorial Board member of 25 international journals and acts as reviewer for more than 80 prestigious Web of Science journals. In addition, he has also published as editor (and co-editor) more than 100 books and as author (and co-author) more than 10 books, 80 book chapters and 400 articles in journals and conferences (more than 200 articles in journals indexed in Web of Science core collection/h-index 45+/6500+ citations and SCOPUS/h-index 53+/8500+ citations).

https://doi.org/10.1515/9783110352986-202

List of contributing authors

Mohammed Mehdhar Al-Mehdhar
Department of Mechanical Engineering
King Fahd University of Petroleum and Minerals
Dhahran 31261, Saudi Arabia

S. Basavarajappa
Indian Institute of Information Technology
Dharwad, Karnataka, India

Jayashree Bijwe
Industrial Tribology, Machine Dynamics and Maintenance Engineering Centre (ITMMEC)
Indian Institute of Technology Delhi
Hauz Khas, New Delhi 110016, India
jbijwe@gmail.com

Satheesan Bobby
Department of Mechanical Engineering
King Fahd University of Petroleum and Minerals
Dhahran 31261, Saudi Arabia

J. Paulo Davim
Department of Mechanical Engineering,
University of Aveiro
3810-193 Aveiro, Portugal
pdavim@ua.pt

Zhiqin Ding
Ministry of Education Key Laboratory of Synthetic and Natural Functional Molecular Chemistry
School of Chemistry and Materials Science
Northwest University
Xi'an 710127, China

G. Gautam
Department of Metallurgical and Materials Engineering
IIT Roorkee
Uttarakhand 247667, India
gauravgautamm1988@gmail.com

Ana Horovistiz
TEMA, Department of Mechanical Engineering,
University of Aveiro
3810-193 Aveiro, Portugal
horovistiz@gmail.com

Kiran T.S.
Department of Mechanical Engineering
Kalpataru Institute of Technology
Tiptur, Karnataka, India
kirants111@gmail.com

N. Kumar
Department of Mechanical Engineering
BIET, Jhansi
Uttar Pradesh 284128, India

Susana Laranjeira
Department of Mechanical Engineering,
University of Aveiro
3810-193 Aveiro, Portugal

Jinjun Lu
Ministry of Education Key Laboratory of Synthetic and Natural Functional Molecular Chemistry
School of Chemistry and Materials Science
Northwest University
Xi'an 710127, China
jjlu@nwu.edu.cn

Umesh Marathe
Industrial Tribology, Machine Dynamics and Maintenance Engineering Centre (ITMMEC)
Indian Institute of Technology Delhi
Hauz Khas, New Delhi 110016, India

Junhu Meng
State Key Laboratory of Solid Lubrication
Lanzhou Institute of Chemical Physics
Chinese Academy of Sciences
Lanzhou 730000, China

https://doi.org/10.1515/9783110352986-203

A. Mohan
Department of Physics
IIT(BHU) Varanasi
Uttar Pradesh 221005, India

S. Mohan
Department of Metallurgical Engineering
IIT(BHU) Varanasi
Uttar Pradesh 221005, India

Meghashree Padhan
Industrial Tribology, Machine Dynamics and Maintenance Engineering Centre (ITMMEC)
Indian Institute of Technology Delhi
Hauz Khas, New Delhi 110016, India

Prasanna Kumar M.
Department of Studies in Mechanical Engineering, University BDT College of Engineering
Davangere, Karnataka, India

Mohammed Abdul Samad
Department of Mechanical Engineering
King Fahd University of Petroleum and Minerals
Dhahran 31261, Saudi Arabia
samad@kfupm.edu.sa

Viswanatha B. M.
Department of Mechanical Engineering
Kalpataru Institute of Technology
Tiptur, India

Xiaoqin Wen
Ministry of Education Key Laboratory of Synthetic and Natural Functional Molecular Chemistry
School of Chemistry and Materials Science
Northwest University
Xi'an 710127, China

Ruiqing Yao
Ministry of Education Key Laboratory of Synthetic and Natural Functional Molecular Chemistry
School of Chemistry and Materials Science
Northwest University
Xi'an 710127, China

Feiyan Yuwen
Ministry of Education Key Laboratory of Synthetic and Natural Functional Molecular Chemistry
School of Chemistry and Materials Science
Northwest University
Xi'an 710127, China

Umesh Marathe, Meghashree Padhan, and Jayashree Bijwe

1 Tribology of carbon fabric-reinforced thermoplastic composites

1.1 Introduction

The chapter focusses on the tribological exploration of carbon fabric (CF)-reinforced thermoplastic polymer (TP) composites. It starts with the diverse demands of friction and wear in various applications and concentrates on the type of polymer composites with special reference to the fabric-reinforced ones including their advantages and limitations. The chapter is split into two parts: the first part focusses on the work done in the authors' laboratory while the other part focusses on the recent efforts reported by other researchers. The chapter depicts the state of the art on the influence of various parameters such as type of fabric, its amount, weave, orientation with respect to the sliding direction, processing technique, quality of fiber–matrix interface, molecular weight (MW) of a polymer and so on, and their bearing on the performance properties such as mechanical strength and tribo-performance in adhesive wear mode. It finally sums up with some salient observations and comments on necessity of advancing the research in some important directions.

1.1.1 Polymer composites in tribology

Tribology is a science of surfaces sliding against each other in a relative motion and encompasses various phenomena such as friction, wear and lubrication [1]. Tribological situations demand all types of combinations of magnitudes of friction and wear (Figure 1.1). Managing these two parameters in right amount is the most challenging task for the researchers in tribology.

There are several factors such as operating conditions, environment and material properties of interacting surfaces that control friction and wear. Wear modes are usually classified into abrasive, adhesive, fatigue, fretting, erosive and corrosive. These occur seldom in isolation and generally in combination. Tribology is also called as a science of making films and breaking films. Adhesive wear is initiated by the construction and destruction of interfacial adhesive junctions by the asperities. For controlling adhesive wear, it is necessary to understand the contributing factors. For polymers and composites, it mainly depends on

- how quickly a beneficial film is formed on the counterface so that metal–polymer interaction will be drastically reduced;
- how firmly it adheres to the counterface so that final interaction of a tribo-couple gets transformed from polymer against metal into polymer/solid lubricants

https://doi.org/10.1515/9783110352986-001

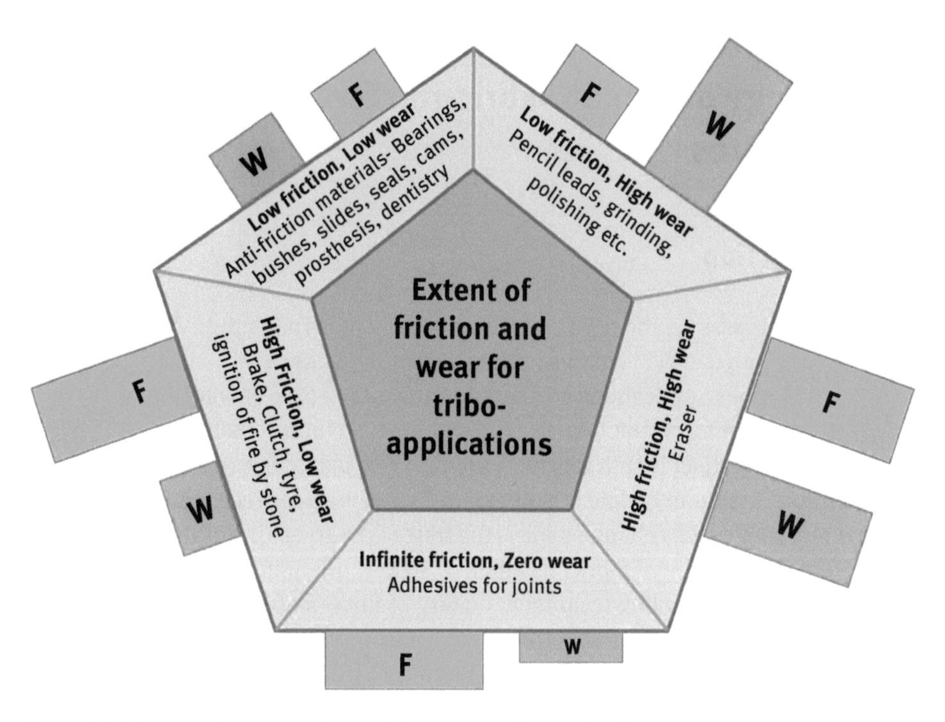

Figure 1.1: Friction and wear requirements for various tribo-applications.

against polymer (since adhesion is minimal, coefficient of friction is very low and wear is also very low); and
- the quality of film transferred (a thin, coherent and uniform film is beneficial, which avoids stick-slip problems also).

Tribo-materials are mainly of three types, namely, metallic, ceramic and polymeric. Each type has its own advantages and limitations. Polymers and composites form a special class because of their unique properties such as
- Self-lubricity (capability to run without any external conventional lubricants such as oils and greases; all polymers do not have self-lubricity. In fact some such rubbers have high friction coefficient – μ)
- Resistance to wear (for most of the polymers excluding poly(tetrafluoroethylene) [PTFE])
- Resistance to corrosion due to high chemical stability and inertness toward the environment
- Resistance to impact
- High compliance and tendency for damping vibration and noise leading to the quiet operation
- Light weight (high specific strength especially for fiber-reinforced polymers [FRPs])

- Ease of production at higher rate even for most complicated components by injection molding without the requirement of machining, polishing and so on leading to low cost of the items

Polymers, however, have some inherent limitations as engineering tribo-materials. Table 1.1 summarizes some of them along with the remedies.

Table 1.1: Limitations and remedies of polymers as tribo-materials.

Limitations	Remedies
Poor thermal stability leading to either degradation or melting at elevated temperature, which finally leads to loss in mechanical and physical properties	Use of specialty polymers such as PI, PTFE, PEI, PES, PEEK* that have higher utility temperature (up to 300 °C). However, for applications beyond this range, polymers are not used.
Low load-bearing capacity, poor strength, modulus and higher tendency to creep limiting them as engineering materials	Inclusion of fillers or fibers (metallic, ceramic or organic) to overcome these limitations to the extent possible.
Very high thermal expansion coefficient, creating problems in clearance	Use of metallic powders or fibers in adequate amount as fillers.
Low TC and dissipativity (frictional heat generated gets accumulated on the surface causing degradation in mechanical, physical and chemical properties of the polymers)	Use adequate amount of metallic contents for increasing the TC or tailoring of the surface with right combination of fillers.
High wear rate of some polymers	Add proper reinforcement, solid lubricants, etc.

Note: PI, poly(imide); PTFE, poly(tetrafluoroethylene); PEI, poly(etherimide); PES, poly(ethersulfone); PEEK, poly(etheretherketone); TC, thermal conductivity.

Hence polymers are seldom used in virgin form but in composite form in tribology. A tribo-composite of a polymer generally contains following classes of ingredients:
- Polymer as a matrix: It holds other ingredients firmly so that they can contribute to the function for which they are added. It supports the fibers so that load gets transferred to the fibers.
- Fibers: These act as a reinforcement and improve strength and resistance to wear. If multifunctional, then they may reduce friction and increase thermal conductivity or other desired properties.
- Solid lubricants: They reduce friction and wear. These are sometimes used in combination synergistically.
- Fillers: They improve thermal conductivity or other favorable functions or act as antioxidants.

Tribo-applications of polymer composites include gears, bearings, slides, tires, shoe soles, automobile brake pads, nonstick frying pans, floorings, chute liners, human joint bearing surfaces and gyroscope gimbals. Typical application areas [3] are given below, where liquid lubrication is not applicable because of

- Possibility of contamination with the products, for example, in industries such as textile, paper, food and pharmaceuticals
- High and low temperatures, where liquid lubricants will either evaporate or solidify
- Space or applications in low pressure/vacuum, where volatility of oils create problems
- Where lubrication is a problem, for example, tribo-components in inaccessible situations, which may be caused by nuclear radiation, chemically hazardous, abrasive or corrosive environment
- Where maintenance is zero, for example, tribo-components in domestic appliances/instruments and toys.
- Where operating conditions are very typical, that is, low sliding speeds, oscillatory motions or frequent starts and stops due to which hydrodynamic lubrication cannot be established by oils.

1.1.2 Types of FRP composites

Fiber reinforcement of fiber-reinforced polymer composites (FRPs) are generally of three types, namely, short fiber-reinforced polymers, continuous (unidirectionally – UD) fiber-reinforced polymers (CFRPs) and fabric-reinforced bidirectionally (BD) reinforced. Table 1.2 describes the major features.

Table 1.2: Fiber reinforcements: types and features.

Reinforcement type	Fibers			
	Short	**UD/continuous**	**BD/fabric (woven–nonwoven)**	**MD**
Aspect ratio	Aspect ratio (l/Φ) is low~ 100–200; hence, lowest strength; probability of defects decreases significantly	High aspect ratio >1,000, high strength, long fibers or tapes; hence, free from major flaw of poor wetting by matrix at crossover points, crimps, etc.	Can be with woven (fabric) or nonwoven fibers, high aspect ratio >1,000; high strength in two directions	High aspect ratio >1,000 (continuous reinforcement, mostly isotropic in nature)
Processability	Easy	Difficult	Easy	Very difficult
Handling	Easy	Difficult	Difficult	Difficult
Compression molding	Yes	Yes	Yes	Yes

Note: UD, unidirectional; BD, bidirectional; MD, multidirectional; l, length of fiber; Φ, diameter of fiber.

1.1.2.1 Types of fabric reinforcements for tribo-composites

Fibers are multifunctional. They enhance the mechanical (mainly strength) and tribological (mainly wear resistance) properties of a composite. They can also impart

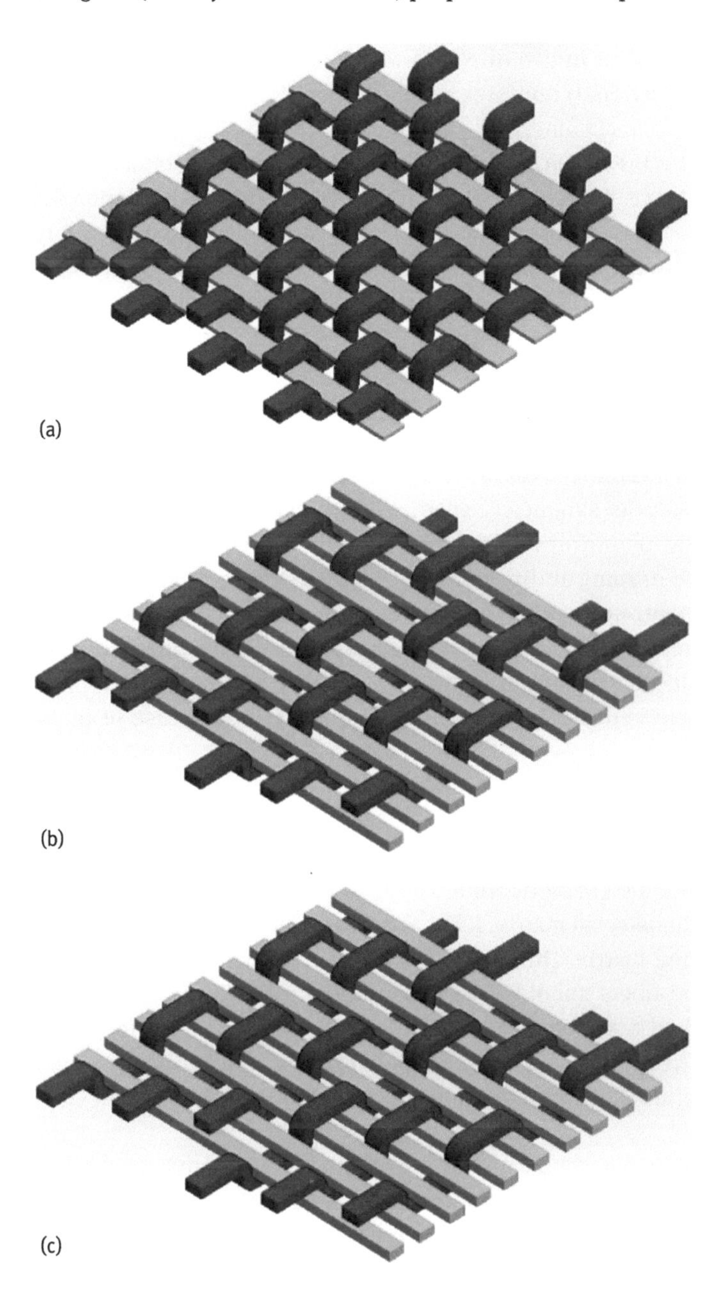

Figure 1.2: Commonly used weaves in fabric-reinforced composites: (a) plain weave (one warp over one weft); (b) twill weave (two warp over two weft); and (c) 4 H satin weave [4].

other properties such as reduction in friction coefficient, increase in thermophysical properties (such as thermal conductivity and expansion coefficient), damping characteristics, impact resistance, modulus and abrasion resistance. Among these, BD reinforcement is most promising and preferred because of very high aspect ratio and hence high reinforcement in two directions, easy processability as compared to UD and MD (multidirectional) reinforcements. Three types of fabrics are generally used for tribo-composites, namely, glass, aramid and carbon/graphite. Among these, glass fabric (GF) is the most cost viable, moderate performer and least beneficial for tribo-performance. Aramid fabric (AF) is moderate for tribo-performance in adhesive wear mode, best in abrasive and erosive wear modes and moderately expensive. CF, on the other hand, is most beneficial for adhesive wear resistance since it imparts highest strength, lubricity and wear resistance, although most expensive. It also imparts thermal conductivity, which is very critical to do away with the frictional heat produced at the interface. It is most preferred in aircraft industries as a structural material. It enjoys a special place in the tribology of FRPCs.

Fabrics vary in types and their weaves (Figure 1.2) and hence properties. Plain weave is well known for its symmetry, good stability, balance and acceptable porosity, whereas it has poor drape, crimp and smoothness. On the other hand, twill weave shows good draping ability, porosity, balance with an acceptable limit of crimp, stability, smoothness and symmetry. But, in contrast to both weaves, satin weave shows an amalgam of excellent draping ability, porosity, crimp and smoothness but lacks severely in symmetry, stability and balance. Hence, handling of satin weave is difficult, whereas it is opposite in the case of plain weave [4].

1.1.2.2 Processing techniques

The first essential requirement of fabric while selecting is about its adequate thermal stability compared to the selected matrix. It has to have higher thermal stability than the melting of a selected matrix. Otherwise, fabric structure will not exist during molding. Moreover, the fibers should have good adhesion for the matrix, which ultimately would lead to a strong fiber–matrix interface and good performance properties including strength. Fibers are generally applied special coatings called sizing during manufacturing, which is generally targeted for typical matrices such as epoxies.

Fabric-reinforced polymer composites (FRPCs) can be fabricated by various methods, but are generally done by compression molding apart from autoclave molding, resin transfer molding and so on. The prepregs are developed by various techniques such as liquid impregnation, solution impregnation (most common and beneficial, if applicable), hand lay-up method, hot melt impregnation, powder prepreg, film stacking, co-mingling/wrapping/co-weaving and resin transfer molding. Each method has some advantages and limitations, which are based on basically selected matrix (thermoset or thermoplastic) and type of fabric selected. TPs, if have solubility in low boiling point and non–corrosive solvents, have wider processing choices.

1.1.2.3 Various factors affecting performance properties of FRPCs

The performance properties of FRPCs depend mainly on the following aspects:
- Type of polymer and its MW
- Type of fabric
- Amount of fabric and polymer
- Weave of fabric
- Fiber–matrix interface
- Orientation of fabric with respect to loading direction
- Other fillers, nanofillers in combination with the fabric
- Processing technique and parameters selected during molding

Each parameter influences all the performance properties to a great extent. Fibers have smooth surfaces and hence fiber–matrix adhesion is less leading to weaker composites. Fibers of carbon, graphite and so on are known for having least interaction with the matrix. If the fiber surface is made rougher, mechanical keying of molten matrix will be promoted to a large extent. If some functional groups such as ketone (–CO), alcohol (–OH) and acid (–COOH) are grown on the fibers by proper treatment, matrix gets covalently bonded and fiber–matrix interface becomes stronger. When both the fabric and filler used are treated and functionalized, it leads to combined beneficial effects. Various methods such as oxidation with nitric acid, $KMnO_4$, $K_2Cr_2O_7$ and others, plasma treatment, gamma irradiation, particles/nanoparticles (NP) of halides of lanthanides and so on, which etch and oxidize the surfaces and impart functional groups, lead to better wetting of the fibers with the matrix. The mechanical strength and tribo-performance improve considerably, and the extent of improvement is a function of type of fiber, matrix, treatment, its dose and so on [5–12].

1.2 Tribology of CF-reinforced polymer composites

This section is divided into two parts. The first part summarizes the investigations from authors' laboratory while the other part focusses on the literature data.

1.2.1 Investigations on tribology of CFRPs (in-house data)

As discussed in Section 1.1.2.3, various parameters such as processing technique; type of polymer; its MW; type, weave, amount and orientation of fabric with respect to loading direction; fiber–matrix bonding; combination of fillers, fibers and so on influence the performance properties significantly including tribological. In spite of this, not many systematic group efforts dealing with the investigations on the influence of these parameters on the tribological behavior are reported. Some amount of work on these aspects is reported in the last two decades [2, 5–23].

1.2.1.1 Influence of fabric: type, weave and orientation

During initial efforts [13, 14] to investigate the influence of type, weave, and orientation of fabric with respect to the sliding direction on mechanical and tribological properties (in abrasive and fretting wear modes), three fabrics of GF (three weaves – plain, twill and woven), AF (plain weave), CF (plain weave) and hybrid fabric (alternate sequence of CF and AF) were selected as a reinforcement (75 wt.%) in poly(etherimide) (PEI, ULTEM 1000) matrix. The prepregs based on either impregnation or hand layup technique were compression molded at 380 °C. The technique is described in Figure 1.3.

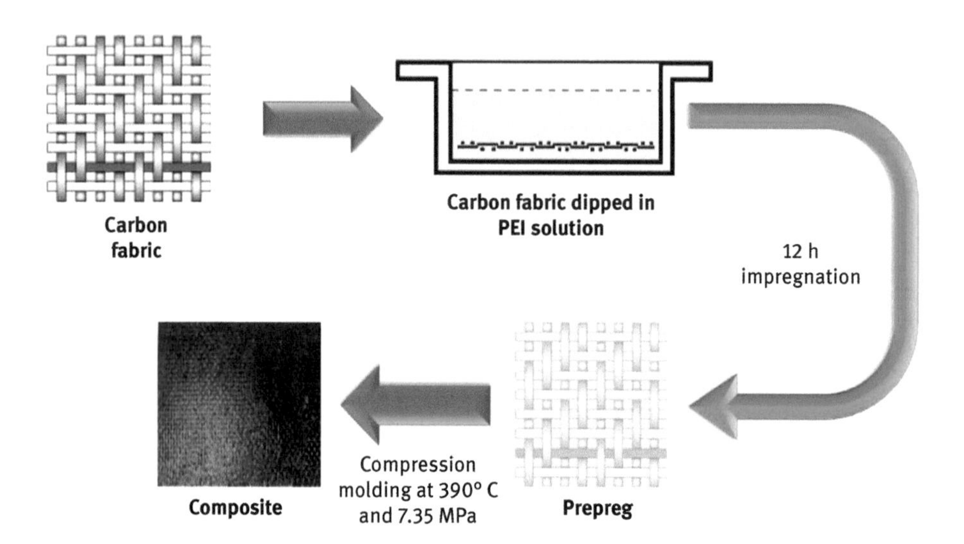

Figure 1.3: Impregnation method for PEI–CF composites [21].

The mechanical and tribological properties of hand layup technique proved inferior to the impregnation (I) technique. Hence, in-depth studies were carried out on the composites with impregnation technique. During comparative studies in identical conditions, CF performed best to impart excellent mechanical properties (tensile, flexural and interlaminar shear strength [ILSS]) while AF proved best only in terms of impact strength. GF proved poorest. CF rendered highest wear resistance (W_R) in fretting wear mode followed by AF and then GF. Since the CF proved best it was selected for further studies as a reinforcement for PEI matrix.

Influence of amount of CF

Contents of CF varied from 40 to 85 wt.% in PEI matrix and details of developed composites are given in Table 1.3.

The major themes for investigations were:

- Amount of fabric – wt.% (IP_{40}, IP_{55}, IP_{65}, IP_{75} and IP_{85} – identical impregnation technique and plain weave)
- Weave – plain, twill and satin (IP_{55}, IT_{55} and IS_{55} – identical impregnation technique and amount of fabric 55%)
- Weave – plain, twill and satin (FP_{52}, FT_{52} and FS_{52} – identical film technique and amount of fabric 52%)

Table 1.3: Details of developed composites [15].

Sl. no.	Amount of fabric	Influence of weave with I technique	Influence of weave with F technique
1	IP_{40}	IP_{55}	FP_{52}
2	IP_{55}	IT_{55}	FT_{52}
3	IP_{65}	IS_{55}	FS_{52}
4	IP_{75}	–	–
5	IP_{85}	–	–

Note: I, impregnation technique; F, film technique; P, plain; T, twill; S, satin; subscripts indicate the content of fabric (by wt.%).

Composites were tribo-evaluated in pin-on-disk configuration where pin of composite (10 mm × 10 mm × 4 mm) slides against mild steel disk (R_a: 0.1–0.2 μm). The speed (1 m/s), sliding duration (2 h) and distance (7.536 km) were kept constant. The load was a variable parameter (200–600 N in steps on 100 N).

Table 1.4 indicates the relative enhancement factor (REF) for mechanical properties for the above-mentioned composites:

$$\text{REF} = \text{Property of a composite/property of a virgin polymer}$$

Figure 1.4 shows the variation of tribological properties as a function of fabric content.

Amount of CF in the range of 65–75 wt.% proved most appropriate for best combination of strength and modulus properties.

Table 1.4: Mechanical properties: relative enhancement factor for composites [15].

Composites	IP_{40}	IP_{55}	IP_{65}	IP_{75}	IP_{85}	IP_{55}	IT_{55}	IS_{55}	FP_{52}	FT_{52}	FS_{52}
T.S	3.14	5.09	6.64	6.58	5.35	5.09	**8.45**	5.47	4.48	4.09	3.14
T.M	18.0	24.33	29.0	28.33	25.3	24.33	**35.33**	25.33	24.33	23.0	17.66
F.S	3.36	3.92	5.45	**6.75**	2.45	3.92	6.34	5.55	1.80	1.63	2.22
F.M	8.78	12.12	15.1	**16.96**	6.60	12.12	16.36	13.93	12.72	8.78	15.75

Note: I, impregnation technique; F, film technique; P, plain; T, twill; S, satin.
Interlaminar shear strength and toughness values were not calculated since data on poly(etherimide) were not applicable/available. The values in bold show the highest achievements.

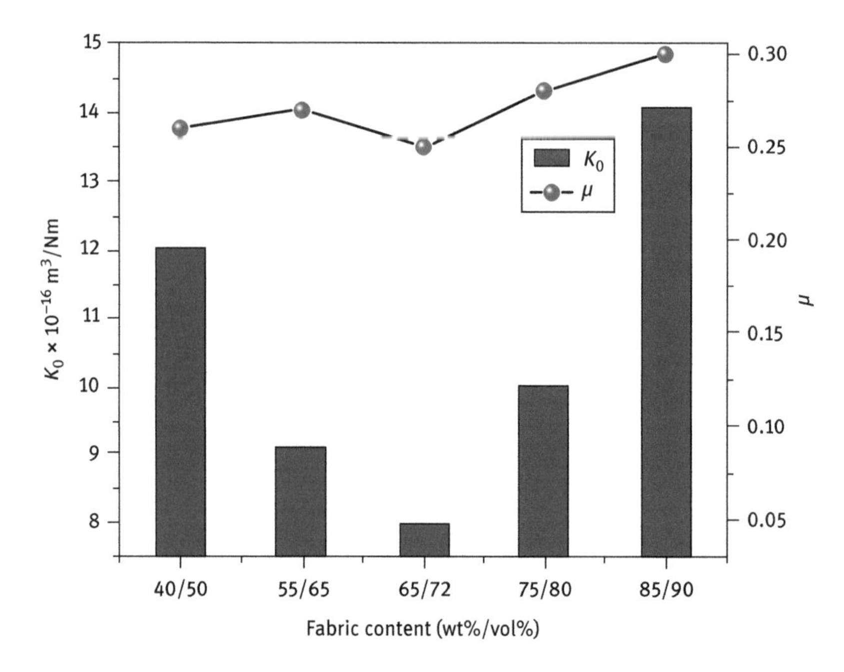

Figure 1.4: Specific wear rate (K_0) and coefficient of friction (μ) as a function of % fabric content for PEI/CF composites at 100 N, 1 m/s [16].

The performance ranking of the composites based on the percentage of fabric for low friction and low wear rate was as follows. For high wear resistance (W_R) and low μ, wt.% of fabric in PEI is:

$$65 >> 55 > 75 >> 40 >> 85.$$

Among the composites, limiting loads (tribo-utility) followed the order (fabric wt.%):

$$IP_{55} = IP_{40}\ (600\ N) > IP_{65} = IP_{75}\ (500\ N) > IP_{85}\ (400\ N).$$

A fairly good wear–property correlation, as shown in Figure 1.5, was observed between K_0 and the combination of mechanical properties such as ILSS and ultimate tensile strength (TS).

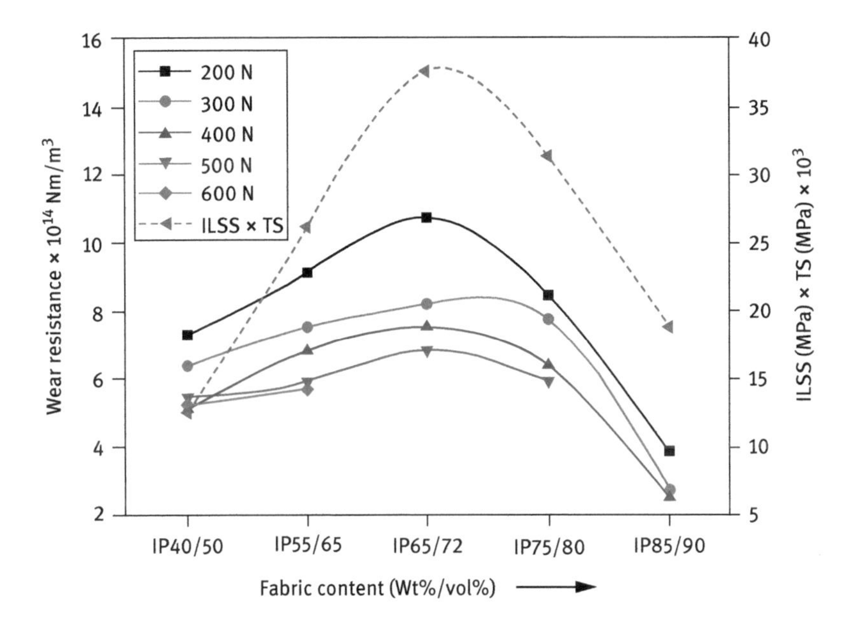

Figure 1.5: Correlation between wear resistance and combination of tensile strength (TS) and interlaminar shear strength (ILSS) of the composites [15].

The fiber–matrix interfacial bonding is at its peak when there is enough amount of matrix to wet the strands, seep into the fabric and then hold the fibers together. The strong interface bonding is reflected in a high ILSS, TS and flexural strength. Twill weave proved best performing. Film (F) technique proved poorest because of nonwetting of crossover points. Solution impregnation (I) technique proved most promising.

Influence of weave of CF

Figure 1.6 shows the details of influence of weave on tribo-performance. W_R significantly depended on a weave while μ was affected marginally. Twill weave showed correlation with mechanical and tribological properties while the others did not show any correlation. The performance order was as follows:

$$\text{Performance (high } W_R \text{ and low } \mu) - \text{Twill} > \text{Plain} > \text{Satin.}$$

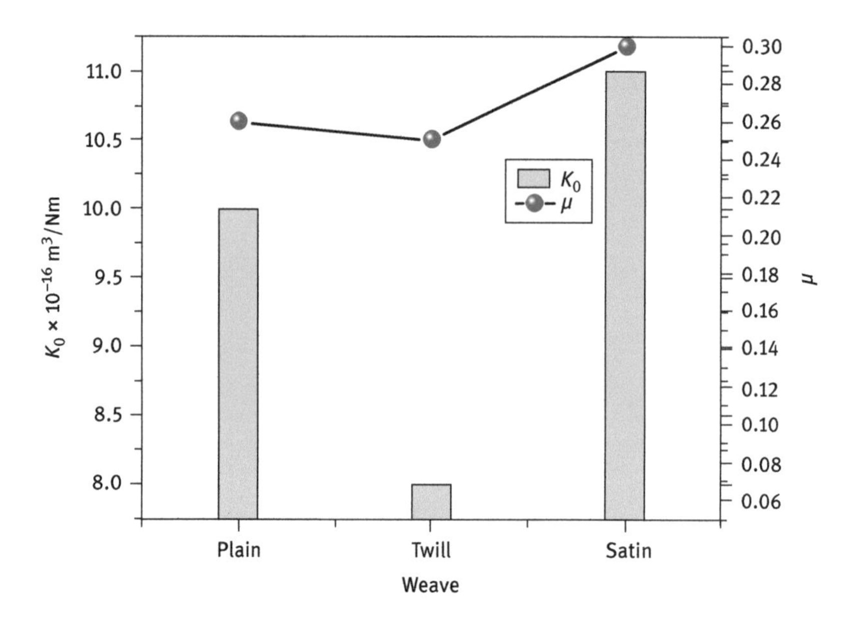

Figure 1.6: Specific wear rate (K_0) and coefficient of friction (μ) as a function of weave of fabric. Adhesive wear mode at 200 N, 1 m/s with counterface as mild steel disk (R_a – 0.2–0.3 μm) for PEI/CF composites [17].

Table 1.5: RWRE of Series I and II composites at various loads [15].

	Series I – variation in amount					Series II – variation in weave		
Composites (N)	**IP_{40}**	**IP_{55}**	**IP_{65}**	**IP_{75}**	**IP_{85}**	**IP_{55}**	**IT_{55}**	**IS_{55}**
200	1.37	1.58	**1.74**	1.50	1	1.1	**1.46**	1
300	1.50	1.65	**1.75**	1.67	1	1.09	**1.42**	1
400	1.48	1.63	**1.76**	1.56	1	1.05	**1.21**	1
500	1	1.05	**1.15**	1.03	–	1.04	**1.05**	1
600	1	1.04	–	–	–	–	–	–

Note: Poorest composite up to 400 N was IP_{85}. For 500 and 600 N was IP_{40}. The values in bold show the highest value.

The factor "relative wear resistance enhancement (RWRE)" that quantified the capability of the composite to enhance the W_R as compared to the poorest is defined as follows:

$$\text{RWRE} = \frac{K_0 \text{ of poorest composite}}{K_0 \text{ of selected composite}}$$

Overall, the weave performance was as follows:
- Twill > Plain > Satin

Thus CF proved to exhibit significant potential to enhance the mechanical and tribological properties (limiting load, wear resistance and μ) in adhesive wear mode. Moreover, the impregnation processing technique proved superior to film technique for enhancing performance properties. Overall, the composite with CF (65 wt.%) showed best strength properties and also enhanced tribo-behavior. Table 1.5 indicates the RWRE of composites for varying fabric content and weaves.

Fabric orientation studies

The orientation of fibers (warp and weft) with respect to sliding direction and orientation of fabric with respect to sliding plane also proved important for controlling the performance of composites.

When the fabric in composite was normal to the sliding plane (50% of fibers in the direction perpendicular to the plane), μ was excessively high (>1.1) showing continuous increase in squeal, vibrations in machine and disk temperature even at the lowest load. So this orientation proved to be forbidden for achieving low friction and low wear. When fabric in the composites was parallel to the sliding plane, composite worked very smoothly. The fabric contained PAN-based carbon fibers, which are known to exhibit very high μ since the graphenes on cylindrical surfaces of fibers are perpendicular to the surface instead of being parallel.

Thus for fabrication of bearings or sliding parts (that are prone to adhesive wear) based on BD composites, following guidelines are recommended for screening the materials [15]:
- Amount of fabric in the composites should be in a moderate range of ≈65 wt.%.
- Impregnation technique is a better choice for fabrication of composites rather than film technique.
- Twill weave should be used for fabrication of composites to attain high $W_{R,}$ low μ and high strength.
- For best results, CF (PAN based) in a composite should be parallel to the sliding plane.

Fiber–matrix interface

In the subsequent work [5–12], the theme was to improve the fiber–matrix adhesion in PEI–CF composites. Based on earlier work, impregnation technique was selected and 67–70 wt.% of fabric of twill weave was selected. The orientation of fabric in composite was parallel to the sliding plane.

The CF was treated with four types of techniques with varying doses as follows:
- Oxidation by HNO_3 (treatment duration: 15–180 min)
- Cold remote nitrogen oxygen plasma (CRNOP) (0, 0.5, 1% O_2 in N_2 plasma)

- Gamma ray irradiation (100, 200, 300 kGy)
- NPs of YbF_3 (0.1, 0.3, 0.5 wt.%)

Each method was expected to enhance the fiber–matrix bonding and hence performance by either growing functional groups as a result of oxidation or by roughening the fiber surface by etching to enhance mechanical keying of matrix with fibers. The oxidation treatment works in two opposite directions. It imparts functional groups on fibers, which interact with the matrix and strengthen the interface. Since it corrodes/etches the smooth surface of a fiber, more surface area is created and extent of interaction with matrix increases by mechanical keying phenomenon. However, these beneficial effects are at the cost of some deterioration in the original strength of a fiber. The composite with treated fibers thus enjoys stronger

Table 1.6: Effect of surface treatment on mechanical properties of fibers and composites [12].

Type of treatment	Fiber/ composite	Fiber properties		Properties of composites			
		AT (% gain in wt.)	FT (load in N)	TS (MPa)	TM (GPa)	FS (MPa)	FM (GPa)
Untreated	F_0/C_0	51.96	246	757	66	927	59
HNO$_3$	F_{15}/C_{15}	54.66	–	781	67	1,003	60
	F_{30}/C_{30}	61.08	–	813	69	1,051	62
	F_{60}/C_{60}	68.05	224	870	72.5	1,118	63.5
	F_{90}/C_{90}	71.58	207	842	70	1,198	65
	F_{120}/C_{120}	71.03	182	778	66	1,136	64
	F_{150}/C_{150}	73.93	–	738	65	1,048	59
	F_{180}/C_{180}	74.5	148	625	63	947	57
Plasma	F_{P1}/C_{P1}	55.9	239	750	67	991	61
	F_{P2}/C_{P2}	63	234	810	72.5	1,134	65
	F_{P3}/C_{P3}	64.88	220	893	79	1,174	73
Gamma	F_{G1}/C_{G1}	56.79	230	–	–	–	–
	F_{G2}/C_{G2}	61.07	217	–	–	–	–
	F_{G3}/C_{G3}	63.83	202	–	–	–	–
YbF_3	F_{Y1}/C_{Y1}	55.5	234	–	–	–	–
	F_{Y3}/C_{Y3}	61.22	227	–	–	–	–
	F_{Y5}/C_{Y5}	62.5	211	–	–	–	–

Note: F and C are used for fibers and composites, respectively, AT, adhesion test; FT, fiber tow tension test; subscripts 0, 15, 30, 60, 90, 120 and 180 indicate treatment time in minutes with HNO_3; P_1, P_2 and P_3 indicate plasma treatment with 0, 0.5, 1% O_2 in nitrogen plasma; G_1, G_2 and G_3 indicate gamma treatment doses of 100, 200, 300 kGy and Y_1, Y_2, and Y_3 indicate nanoparticle treatment with 0.1, 0.3 and 0.5% YbF_3.

surface but slightly weaker fibers. Hence, the treatment has to be done judiciously. In each method, dose varied to get the optimum value for highest enhancement (Table 1.6).

Significant enhancement in mechanical properties of composites was observed due to the treatment of fabric. Generally, studies revealed that the optimum doses were –0.3 wt.% of NPs of YbF_3, 90 min treatment with HNO_3, 300 kGy of gamma ray

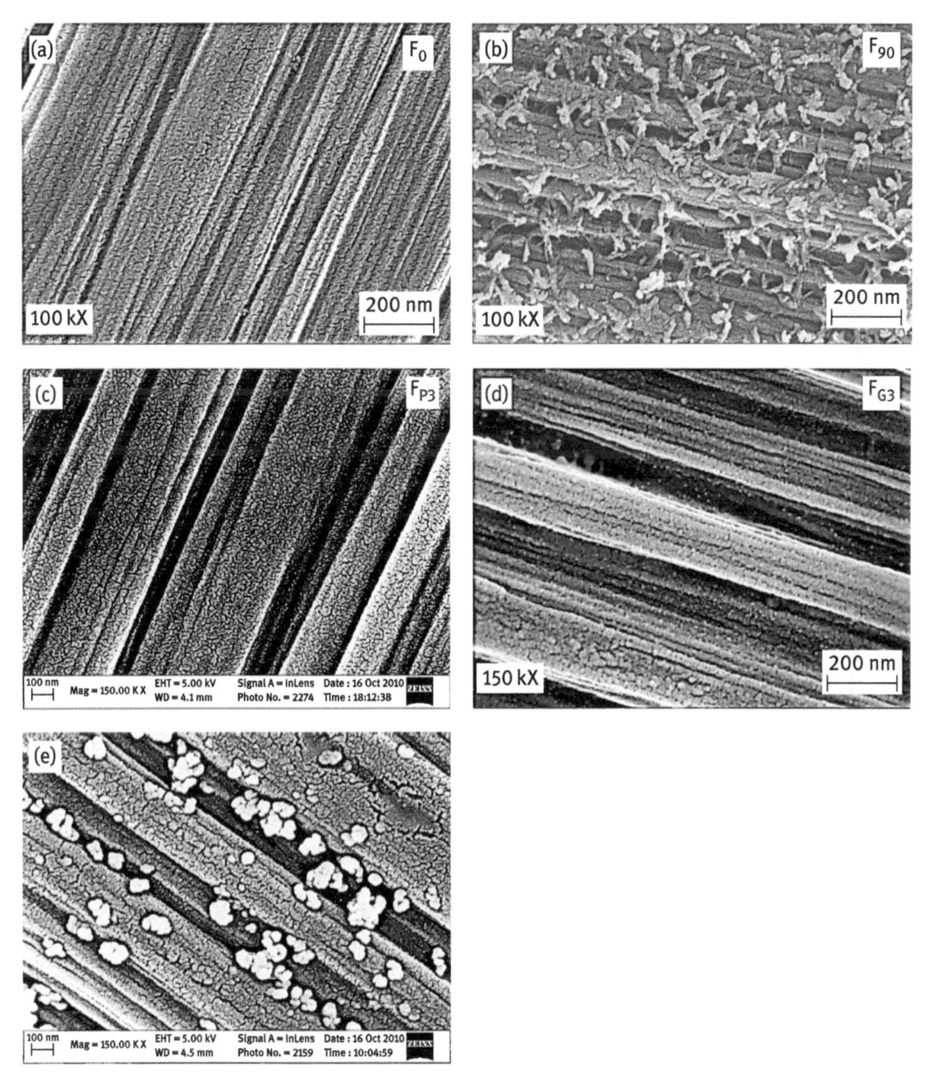

Figure 1.7: Field emission scanning electron microscopy micrographs of treated fibers: (a) F_0 – untreated; (b) F_{90} – treated with HNO_3 for 90 min; (c) F_{P3} – treated with 1% O_2 + N_2 plasma; (d) F_{G3} – treated with 300 kGy of gamma ray irradiation; and (e) FY_3 – treated with 0.3% nano-YbF_3 [5–12] (Copyright permission granted from Elsevier for (c)).

irradiation and 1% O_2 in N_2 plasma treatment. Effectiveness of method was in the order: YbF_3 > HNO_3 >> gamma radiation > CRNOP treatment [12]. Figure 1.7 shows the surfaces of treated fibers with optimum doses.

Adhesive wear was evaluated in a pin (10 mm × 10 mm)-on-disk (mild steel with R_a value of 0.1–0.2 μm) configuration and the fabric in the composite was always parallel to the sliding plane and warp fibers were parallel to the sliding direction. Load was variable parameter (200, 400 and 600 N) and other input parameters were fixed (speed 1 m/s; sliding duration 2 h; sliding distance 7.121 km). The influence of nitric acid treatment on tribo-performance is shown in Figure 1.8. The treatment enhanced tribo-performance apart from mechanical properties. Good correlation was observed with wear resistance and ILSS of the composites.

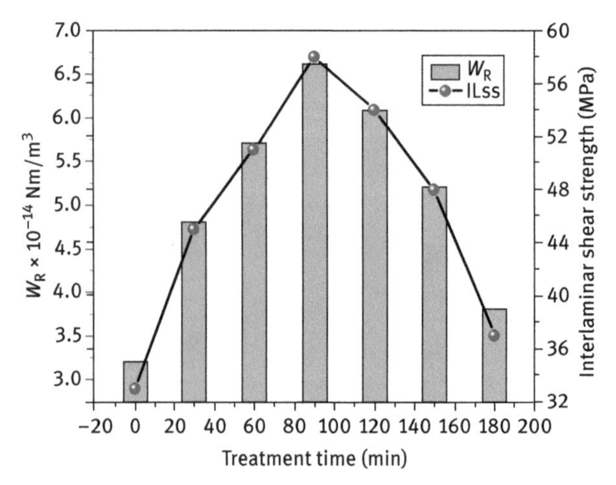

Figure 1.8: W_R and ILSS correlation as a function of treatment time with HNO_3 [12].

Figure 1.9 shows the variation of μ and K_0 of PEI–CF composites as a function of load for different treatment methods applied to the CF. The composites selected were with optimum doses, which showed highest properties and tribo-performance in each series. The wear resistance increased almost two times due to treated fabric and performance order was

$$C_{Y3} > C_{90} > C_{G3} > C_{P1} > C_0$$

(C_0 – composite with untreated fabric)

Thus, NP treatment proved best while plasma treatment proved least beneficial.

Micrographs in Figure 1.10 show crossover points of fibers (without treated 10 (a) and (b) gamma ray treated). The crossover points are the points of weakness when fabric is sheared during sliding. The micrograph (b) clearly shows how the damage to the fibers gets reduced when interface becomes stronger.

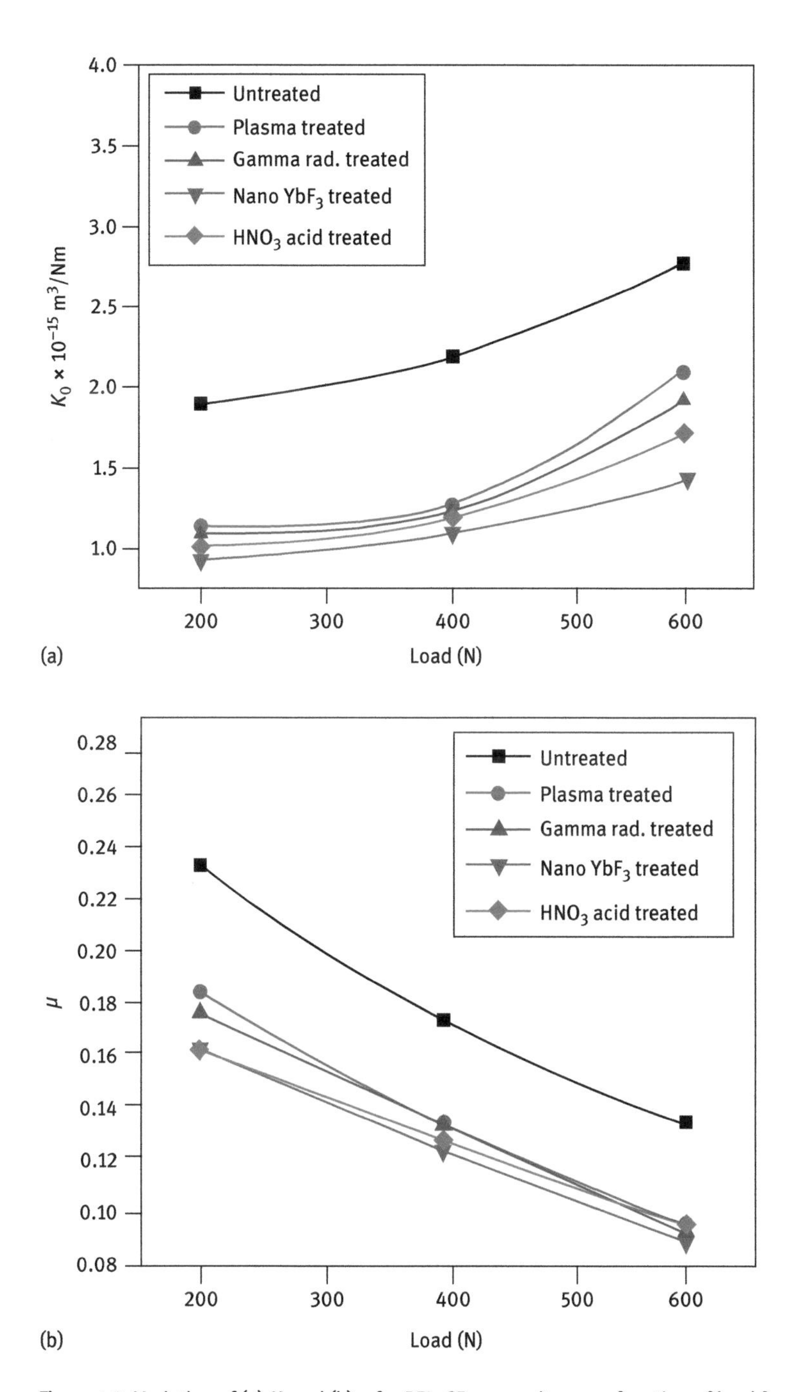

Figure 1.9: Variation of (a) K_0 and (b) μ for PEI–CF composites as a function of load for different surface treatment methods of CF [5–7, 12].

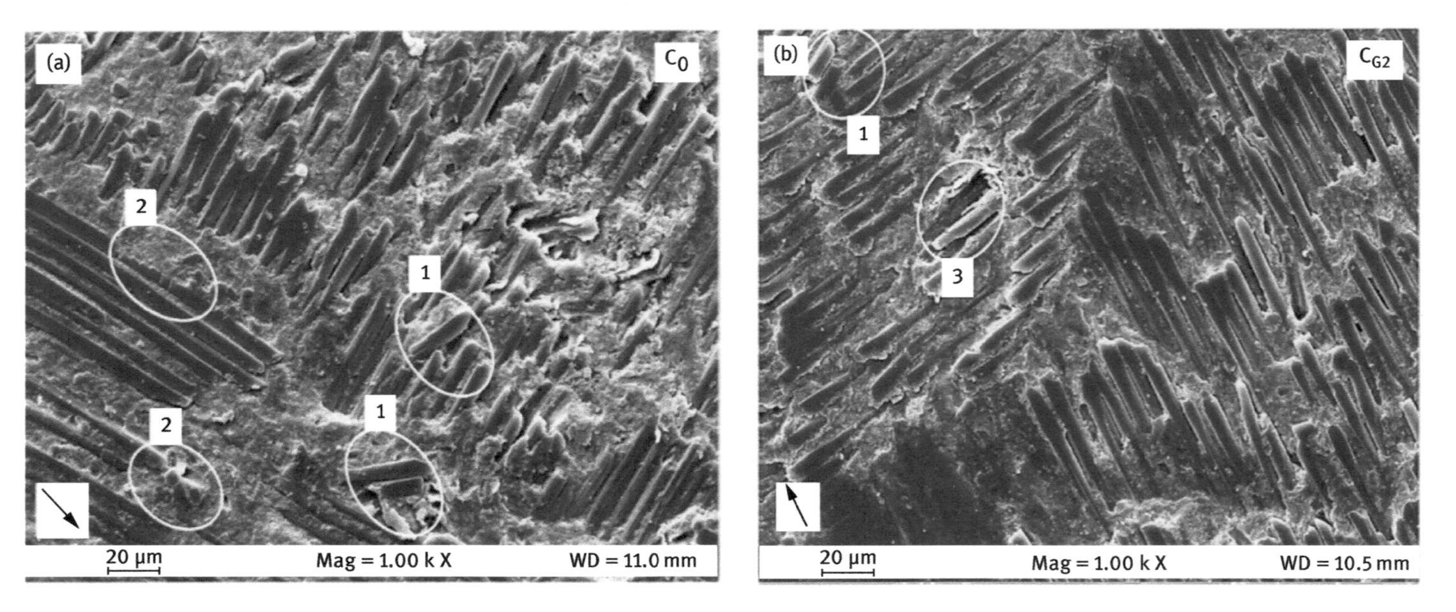

Figure 1.10: Scanning electron microscopic micrographs of worn surfaces of (a) C_0 – untreated and (b) C_{G2} (gamma-irradiated fabric) 200 N; 1 – fiber peeling off from the matrix, 2 – fiber breakage, 3 – fiber–matrix debonding, 4 – back-transferred polymer matrix from the counterface [7]. Copyright permission granted from Elsevier.

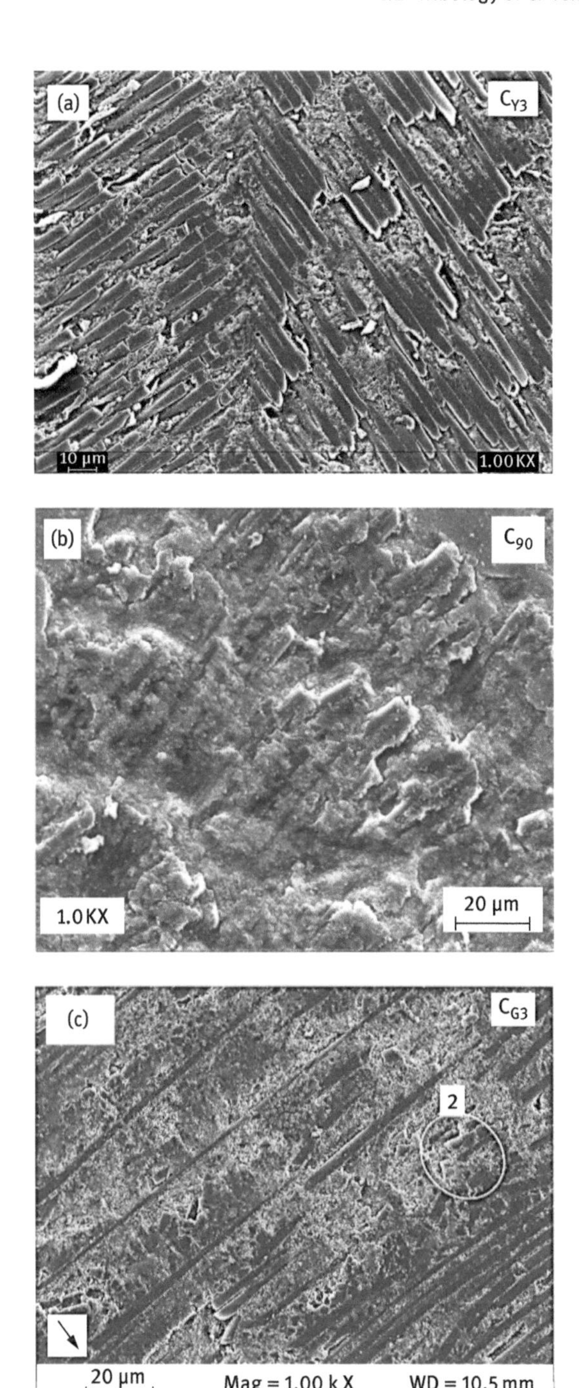

Figure 1.11: Scanning electron microscope micrographs of worn surfaces of composites worn under 200 N load in the order of decreasing wear resistance: (a) C_{Y3} > (b) $C_{90} \geq$ (c) C_{G3} > (d) C_{P3} >> (e) C_0 [5–7, 12]. Copyright permission granted from Elsevier for (c) and (d).

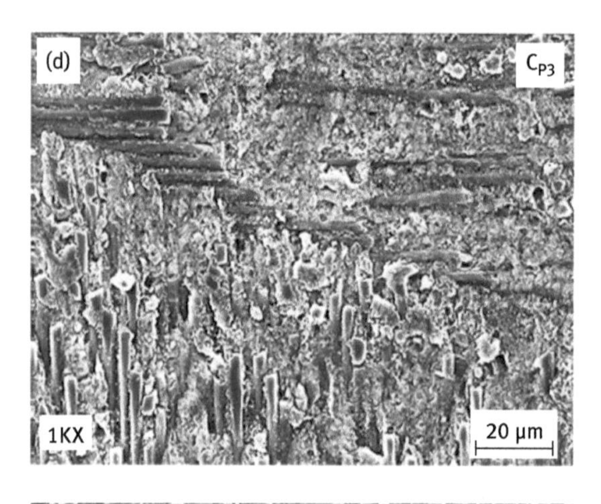

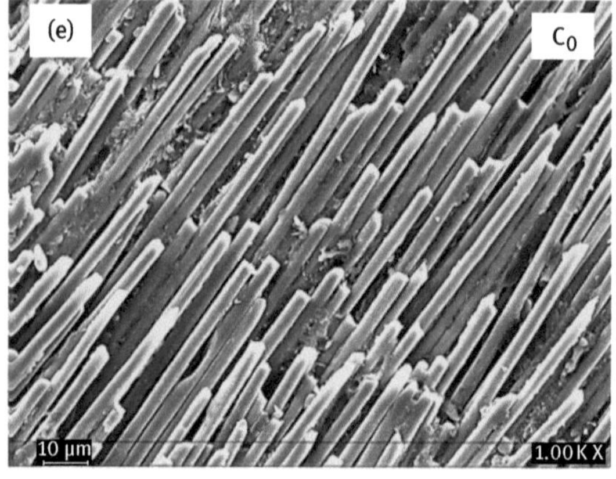

Figure 1.11: (continued)

Figure 1.11 shows the worn surfaces of composites in decreasing order of wear resistance indicating how the difference in surface treatments influence the fiber–matrix interface and hence wear performance.

Influence of MW of a polymer on tribo-performance of composites

In the subsequent work, with a view to study the influence of MW of a polymer on mechanical properties and tribo-performance [18, 20–23], Poly(ethersulfone) (PES) matrix of three types with varying MW designated as L (low MW), M (medium MW) and H (high MW) were selected. The studies were done in the conditions similar to Section "Fabric orientation studies."

Table 1.7 shows the specifications of polymers and their tribo-performance. It was observed that higher the MW, higher was the tribo-performance (low wear rate and μ). Wear rate and μ are $PES_L > PES_M > PES_H$, but performance order is opposite ($PES_H > PES_M > PES_L$), indicating selection of higher MW polymer for better performance.

Table 1.7: Coefficient of friction and specific wear rates for pristine PES (load – 50 N, sliding time – 1 h, sliding speed – 1 m/s) [21].

Polymers	MW	MFI (g/10 min)	Tensile strength (MPa)	$K_0 \times 10^{-13}$ m³/Nm	μ
PES_H	46,020	33	90	6.3	0.29
PES_M	39,825	50	83	6.5	0.30
PES_L	37,770	55	80	8.1	0.33

Note: Subscripts H, M and L indicate high, medium and low MW of poly(ethersulfone) (PES).

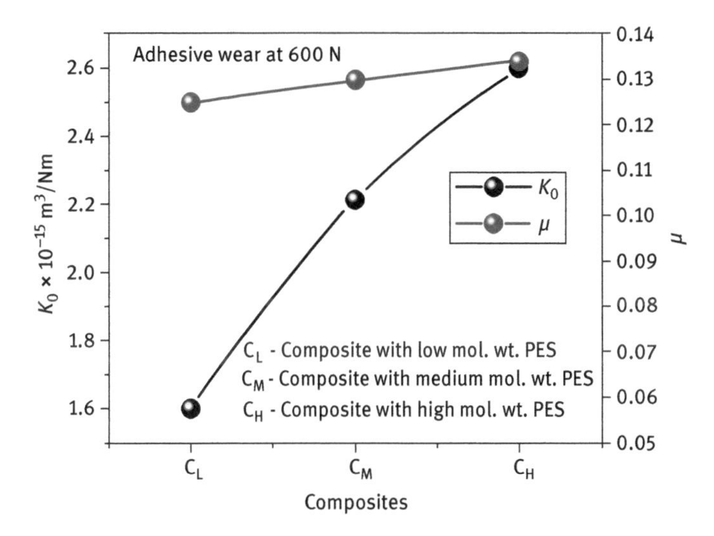

Figure 1.12: Variation in K_0 and μ for PES–CF composites with varying MW of PES [22].

In order to investigate the influence of MW of matrix on performance of composites, twill weave CF fabric (65 wt.%) (CRNOP treated) was selected and three composites designated as C_L (composite with PES of low MW), C_M (composite with PES of medium MW) and C_H (composite with PES of high MW) were fabricated and tribo-evaluated. The results are shown in Figure 1.12.

Interestingly C_L (composite with low MW PES) performed best followed by C_M (composite with moderate MW PES), and C_H (composite with highest MW PES) performed poorest.

Interestingly, these were exactly opposite to that with the pristine polymers. In the case of polymers, higher the MW, lower is the melt flow index, lower is the tendency of the melt to flow, higher is the melt flow viscosity and lower is the wetting capability with the fibers. So finally in the composite, higher the MW of PES, lower was the wettability with fibers and lower was the performance [21]. This was clearly observed in the case of SEM studies (Figure 1.13).

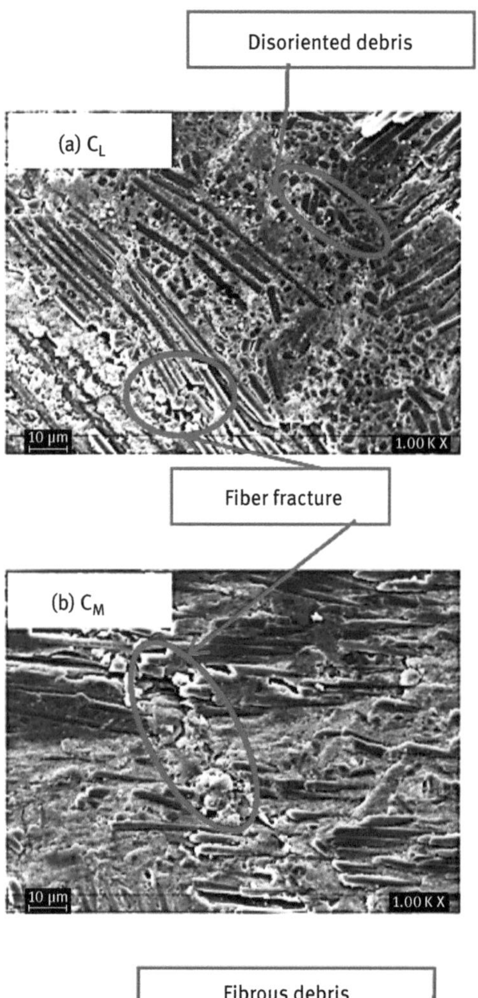

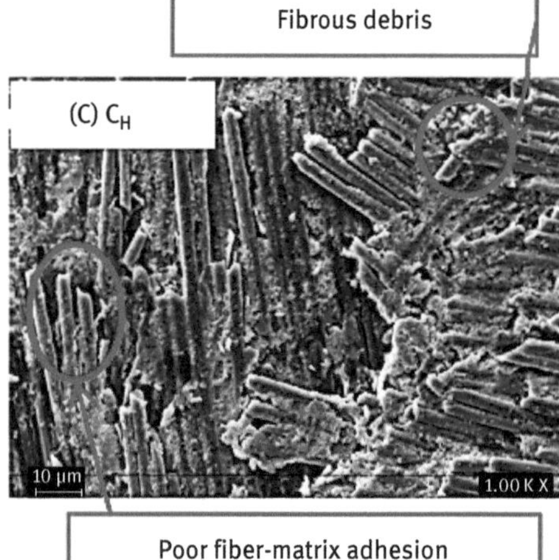

Figure 1.13: SEM micrographs of surfaces of composites sliding under 100 N load and 1 m/s speed: (a) C_L (highest W_R) showing very good fiber–matrix adhesion and minimal damage to fibers, more wear thinning rather than microcutting which was responsible for high W_R; (b) C_M (moderate W_R) showing adequately good fiber–matrix adhesion and pulled out fibers, which was responsible for moderate W_R; (c) C_H (lowest W_R) showing poor fiber–matrix adhesion and excessively pulled out fibers, which was responsible for lowest W_R [22]. Copyright permission granted from Elsevier.

1.2.2 Investigations on tribology of CFRPs – literature data

This section highlights the trends in the literature of tribology of CF-reinforced TP composites carried out by tribologists other than the authors.

Ye and Daghyani [24] studied CF and poly(etheretherketone) (PEEK) commingled fiber composites with 55% (by vol.) of CF in adhesive wear mode. The composites were prepared by heating up to 400 °C and subjected to a pressure of 1 MPa in compression molding. However, the cooling of the composites are classified into three types as follows:

- Cooled in the compression molding machine under pressure (≈3 °C/min)
- Cooled in air under ambient pressure (≈10 °C/min)
- Quenched in engine oil (≈150 °C/min)

Pin-on-disk configuration was used with counterface as mild steel disk (R_a 0.04–0.05 μm) for tribo-testing in three orientations (P, NP, AP). The conditions of tribo-testing are as follows:

- Pressure – 2.1, 3.3, 3.9, 5.2 MPa
- Velocity – 0.985, 1.36 m/s

It was observed that with increase in pressure × velocity, wear depth increased. Effect of orientation on K_0 was in the order: N > AP > P.

Effect of cooling on K_0 was in the order: press cooling > oil quenching > air cooling (for all orientations).

This showed that material possesses different morphology under different cooling rates. High wear rate of air-cooled composites was attributed to poor consolidated material as compared to other types of cooling. It was also observed that the wear rate was significantly affected if the orientation was changed, but the process of cooling had marginal effect.

CF was also explored with PTFE matrix [25, 26]. Liu et al. [25] studied the effect of air plasma, HNO_3 treatment and effect of nano-TiO_2 (2–8%) on CF (50% by vol) – PTFE composites. Composites were fabricated by dispersion impregnation technique followed by hot press molding. Ring on block configuration with 440C steel ring as counterface (R_a – 0.15 μm) was used for tribo-testing with following parameters:

- Load – 300 N
- Speed – 0.2 m/s
- Time – 2 h

It was observed that fiber treatment decreased the μ and K_0 as shown in Figure 1.14.

Order of performance: plasma > HNO_3 > untreated.

It was also observed that inclusion of nano-TiO_2 led to a drastic reduction in μ and K_0 up to 4% with increase in contents. On the basis of the above results, a composite with air plasma-treated CF and 4% nano-TiO_2 was evaluated for tribo-performance to examine the synergistic effect.

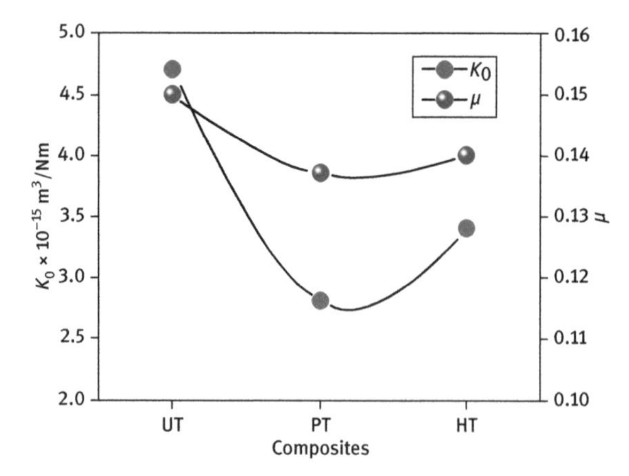

Figure 1.14: Variation of μ and K_0 for untreated (UT), plasma-treated (PT) and HNO_3-treated CF–PTFE composites [25].

Order of performance: plasma + nano-TiO_2 > nano-TiO_2 > plasma > untreated.

Liu et al. [26] also studied the mechanical and tribological properties of CF-reinforced PTFE composites to optimize the CF content. Satin weave of CF was used. CF varied from 30% to 70% by volume. The composites were fabricated by compression process at 380 °C under pressure of 15 MPa. Ring on block configuration with 440C steel ring as counterface (R_a – 0.15 μm) was used for tribo-testing. The operating conditions for tribo-testing are as follows:
- Load – 300 N
- Speed – 0.2 m/s
- Time – 2 h

TS of the composites for varying PTFE content was in the order of

$$40 > 50 > 30$$

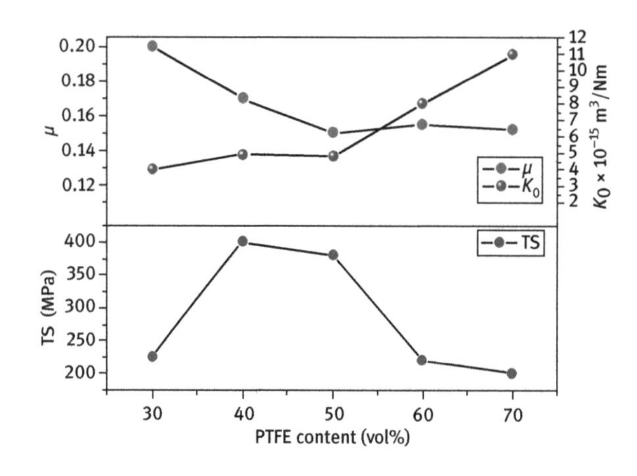

Figure 1.15: Variation of tensile strength and tribo-properties of PTFE-CF composites as a function of PTFE content [26].

It was observed that with increase in load, μ decreased but K_0 increased. Hence, 50% PTFE proved optimum for tribological and mechanical properties as shown in Figure 1.15.

1.3 Tribology of surface-engineered fabric-reinforced polymer composites

While formulating a tribo-composite, solid lubricants are generally added in the polymer composites to reduce friction and wear further. The most common dry solid lubricants are graphite, MoS_2, WS_2 and PTFE. The specialty of such materials is their layer lattice structure (barring PTFE) which offers very low resistance to shearing. These are materials with very low surface energy and transfers film on the counterface very quickly, which is responsible for reduction in friction and wear. However, their low energy and inert nature lead to significant deterioration in mechanical properties of composites. As tribology is the science of surfaces, it is always beneficial to tailor the surfaces rather than bulk. If solid lubricants are placed only on surfaces, it will not only preserve the strength of a composite to a great extent, but will lead to cost saving. Instead of micron-sized particles, nanosized particles or mixed with metallic fillers can also be judiciously placed. Figure 1.16 shows the sketch of such composite.

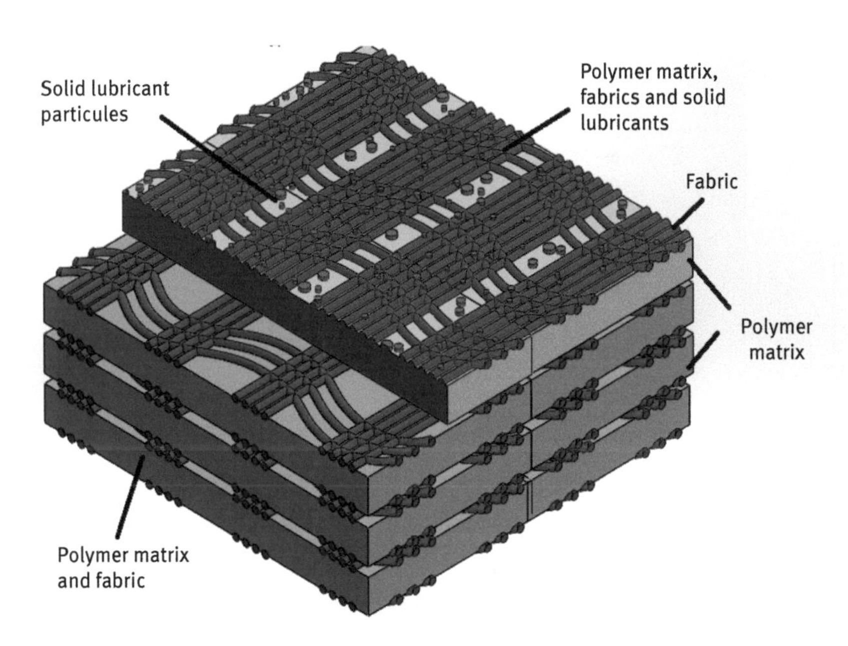

Figure 1.16: Schematic of surface-tailored composite.

Table 1.8: Details of PTFE particles used to tailor the surfaces of composites [21].

Composites designation*	Av. PTFE particle size (FESEM studies) (nm)	Shape of PTFE fillers
C_N	50–80	Highly spherical
C_{SM}	200–250	Subrounded
C_M	400–450	Subangular

Note: *C_N – composite with treated CF and nanosize (50–80 nm) PTFE on the surface.
C_{SM} – composite with treated CF and submicron size (150–200 nm) PTFE on the surface.
C_M – composite with treated CF and micron size (400–450 nm) PTFE on the surface.

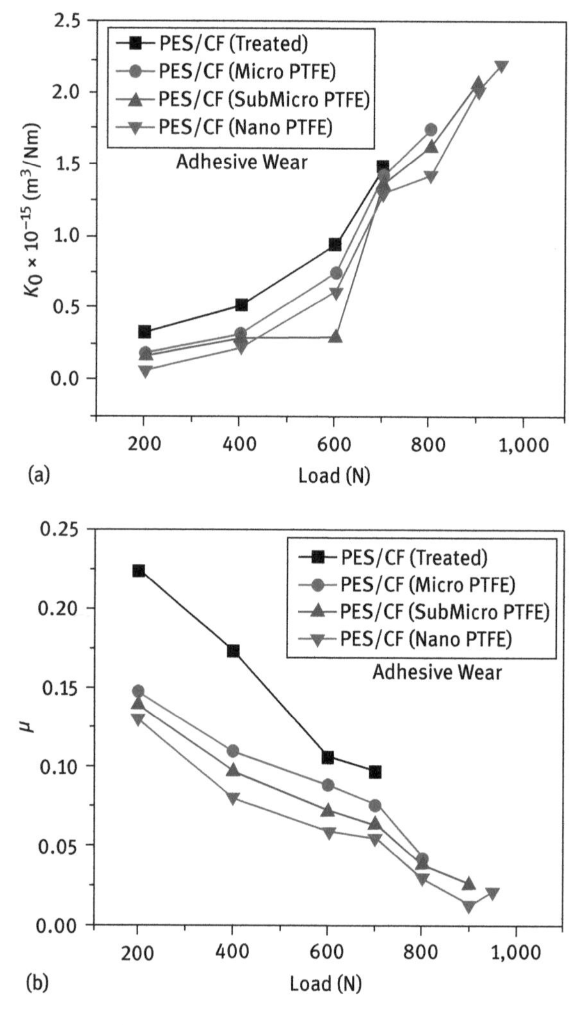

Figure 1.17: (a) Specific wear rates and (b) coefficient of friction as a function of load for surface designed series of composites [20].

Sharma and Bijwe [20] studied the influence of PTFE particles (Table 1.8) on the top three layers on tribo-performance and the essence is shown in Figure 1.17.

Surface designing enhanced the limiting load values of composites significantly from 700 to 950 N, limiting running time from 8 to 21 h; reduction in μ (from 0.12 at 700 N load to 0.065 at 900 N load) and wear rate (from 1.92 to 1.8×10^{-15} m^3/Nm at 700 N load). The increased surface area of contact due to the inclusion of nano-PTFE at the composite surface was responsible for enhanced tribo-performance of C_N composite. Figure 1.17 describes the effect of size of particles on the surfaces on friction and wear of CF-reinforced PES composites.

1.4 Summary

Tribology of fabric-reinforced thermoplastic-based composites is not yet given due attention by the researchers when compared to their short fiber-reinforced polymer composites (SFRPCs), which could be because of complexity involved in the processing of the composites. Injection molding of SFRCs is very easy and facilities are easily available.

The major observation could be about nonavailability of systematic efforts for investigating various aspects such as influence of type of matrix and fabric, their amount, weave and orientation, in case of high-performance specialty polymers such as PEEK and Poly (aryl ether ketone) (PAEK). This could be because of problems associated with processing. Such composites cannot be processed by impregnation method since polymers do not have appropriate solvents. Other techniques such as film technique or powder sprinkling do not give good quality tribo-composites since crossover points are not taken care of. This calls for novel methods of processing for such composites.

In spite of the fact that surface engineering is the smart technique for manipulating strength, cost and tribo-performance of a composite, very few efforts are focused and calls for attention.

The nanofillers are known to have exhibited significant potential in case of SFRPCs. However, they are hardly explored in the case of CFRP composites to enhance their performance to a greater extent.

References

[1] Bhushan B. Introduction to tribology, revised. New York: John Wiley & Sons, 2013.

[2] Bijwe J, Sharma M. Nano and Micro PTFE for Surface Lubrication of Carbon Fabric Reinforced Polyethersulphone Composites, in Tribology of Nanocomposites, 1st edition, J. P. Davim, Ed. Springer-Verlag Berlin Heidelberg 2013; 19–39.

[3] Lancaster JK. Dry bearings: a survey of materials and factors affecting their performance. Tribology 1973;6(6):219–251, 1973.

[4] "Woven fabrics." Available at: https://netcomposites.com/guide-tools/guide/reinforcements/woven-fabrics/ Accessed: 5 Feb 2018.
[5] Tiwari S, Bijwe J, Panier S. Enhancing the adhesive wear performance of polyetherimide composites through nano-particle treatment of the carbon fabric. J Mater Sci 2012;47(6):2891–8.
[6] Tiwari S, Bijwe J, Panier S. Adhesive wear performance of polyetherimide composites with plasma treated carbon fabric. Tribol Int 2011;44(7–8):782–8.
[7] Tiwari S, Bijwe J, Panier S. Gamma radiation treatment of carbon fabric to improve the fiber-matrix adhesion and tribo-performance of composites. Wear 2011;271(9–10):2184–92.
[8] Tiwari S, Bijwe J, Panier S. Influence of plasma treatment on carbon fabric for enhancing abrasive wear properties of polyetherimide composites. Tribol Lett 2011;41(1):153–62.
[9] Tiwari S, Bijwe J, Panier S. Tribological studies on polyetherimide composites based on carbon fabric with optimized oxidation treatment. Wear 2011;271(9–10):2252–60.
[10] Tiwari S, Bijwe J, Panier S. Role of nano-YbF 3-treated carbon fabric on improving abrasive wear performance of polyetherimide composites. Tribol Lett 2011;42(3):293–300.
[11] Tiwari S, Bijwe J, Panier S. Polyetherimide composites with gamma irradiated carbon fabric: Studies on abrasive wear. Wear 2011;270(9–10):688–94.
[12] Tiwari S. Role of fabric-matrix Interface on mechanical and tribological properties of carbon fabric – polyetherimide composites, PhD Thesis. Indian Institute of Technology Delhi, 2011.
[13] Indumati J. Friction and wear studies on polyetherimide and composites, PhD Thesis. Indian Institute of Technology Delhi, 2000.
[14] Bijwe J, Indumathi J, Satapathy BK, Ghosh AK. Influence of carbon fabric on fretting wear performance of polyetherimide composite. J Tribol 2001;124(4):834–9.
[15] Rattan R. Investigations on mechanical and tribological properties of carbon fabric reinforced polyetherimide composites, PhD Thesis. Indian Institute of Technology Delhi, 2006.
[16] Bijwe J, Rattan R. Carbon fabric reinforced polyetherimide composites: Optimization of fabric content for best combination of strength and adhesive wear performance. Wear 2007;262:749–58.
[17] Rattan R, Bijwe J, Fahim M. Optimization of weave of carbon fabric for best combination of strength and tribo-performance of polyetherimide composites in adhesive wear mode. Wear 2008;264(1–2):96–105.
[18] Sharma M, Bijwe J, Mitschang P. Wear performance of PEEK-carbon fabric composites with strengthened fiber-matrix interface. Wear 2011;271(9–10):2261–8.
[19] Sharma M, Bijwe J, Singh K. Studies for wear property correlation for carbon fabric-reinforced PES composites. Tribol Lett 2011;43(267):267–73.
[20] Sharma M, Bijwe J. Surface designing of carbon fabric polymer composites with nano and micron sized PTFE particles. J Mater Sci 2012;47(12):4928–35.
[21] Sharma M. Carbon fabric reinforced polymer composites: Development, surface designing by micro and nano PTFE and performance evaluation, Phd Thesis. Indian Institute of Technology Delhi, 2011.
[22] Sharma M, Bijwe J. Influence of molecular weight on performance properties of polyethersulphone and its composites with carbon fabric. Wear 2012;274–275:388–94.
[23] Sharma M, Bijwe J. Influence of fiber-matrix adhesion and operating parameters on sliding wear performance of carbon fabric polyethersulphone composites. Wear 2011;271 (11–12):2919–27.
[24] Ye L, Daghyani HR. Sliding friction and wear of carbon fibre-polyetheretherketon commingled yarn composites against steel. J Mater Sci Lett 1996;15:1536–8.
[25] Liu P, Huang T, Lu R, Li T. Tribological properties of modified carbon fabric/polytetrafluoroethylene composites. Wear 2012;289:17–25.
[26] Liu P, Lu R, Huang T, Wang H, Li T. A Study on the mechanical and tribological properties of carbon fabric/PTFE composites. J Macromol Sci Part B Phys 2012;2348(February 2017):786–97.

Appendix

Abbreviations

μ	Coefficient of friction
Φ	Diameter
4 H	4 Hardness
σε	Ultimate tensile strength × elongation at break
AF	Aramid fabric
AP	Antiparallel
BD	Bidirectional
CF	Carbon fabric
CFRP	Continuous fiber-reinforced polymer
CRNOP	Cold remote nitrogen oxygen plasma
d	Diameter
D	Distance
F	Film technique
FM	Flexural modulus
FRP	Fiber reinforced polymer
FRPC	Fabric reinforced polymer composite
GF	Glass fabric/fiber
H	Hardness
HNO_3	Nitric acid
ILSS	Interlaminar shear strength
I	Impregnation technique
K_0	Specific wear rate
$KMnO_4$	Potassium permanganate
$K_2Cr_2O_7$	Potassium dichromate
l	Length
L	Load
MD	Multidirectional
MFI	Melt flow index
MoS_2	Molybdenum disulfide
MW	Molecular weight
N	Normal
NP	Nanoparticles
P	Plain weave
P-AP	Parallel to plane and antiparallel to sliding direction
PEEK	Poly(etheretherketone)
PEI	Poly(etherimide)
PES	Poly(ethersulfone)
PI	Poly(imide)
PTFE	Poly(tetrafluoroethylene)
R_a	Roughness
REF	Relative enhancement factor
RT	Room temperature
RTM	Resin transfer molding
RWR	Relative wear resistance

RWRE	Relative wear resistance enhancement
S	Satin weave
SEM	Scanning electron microscope
SFRP	Short fiber-reinforced polymer
T	Twill weave
t	Time
TC	Thermal conductivity
TiO_2	Titanium dioxide
TM	Tensile modulus
TP	Thermoplastics
TT	Treated
UD	Unidirectional
UT	Untreated
V	Velocity
VARTM	Vacuum-assisted resin transfer molding
Vol.	Volume
W_R	Wear resistance
WS_2	Tungsten disulfide
Wt.	Weight
YbF_3	Ytterbium trifluoride

Ana Horovistiz, Susana Laranjeira, and J. Paulo Davim

2 Effect of glass fiber reinforcement and the addition of MoS_2 on the tribological behavior of PA66 under dry sliding conditions: A study of distribution of pixel intensity on the counterface

2.1 Introduction

In this work we proposed a new approach of analysis of the uniformity of the transfer film on counterface, in tribological tests, based on a statistical criterion analysis of the distribution of pixel intensity. The relationship between the transfer film on counterface and the tribological behavior of the polyamide 66 (PA66), polyamide 66 filled with molybdenum disulfide (PA66 + MoS_2) and polyamide with 30% of glass fiber reinforcement (PA66GF30) was analyzed. The experiments were carried out on a pin-on-disk tribo-meter, in dry sliding conditions, and the material of counterface disk was a steel Ck45K. The set test conditions were selected based on a pv limiting value for the PA66 sliding against a steel counterface. For these conditions, the material with the best tribological behavior (lower friction coefficient and higher wear resistance) was PA66GF30. The results also suggested that the use of PA66 + MoS_2 is recommended in situations where temperature is a determining factor.

Among the so-called engineering plastics, the polyamides (PA) are gaining more and more industrial applications, due to its good mechanical properties, considerable impact resistance, abrasion and wear resistance [1]. The polarity of the amide functional group, CONH, is responsible for high mechanical strength and good chemical resistance of this polymer class. In addition, the flexibility of the carbon chains confers the characteristics of high lubrication, low friction and good resistance to abrasion to this class of polymers [2]. PA6 and PA66 cover a wide range of properties and are commercially available. This is particularly true of the outstanding tribological properties of PA66, which are attributed to its crystallinity by the polar amide groups bonded by hydrogen bonding [2]. The tribological potential of PA and its composites have been evaluated by many authors in different wear conditions. In a large number of these investigations, the effects of fiber reinforcement and solid lubricant additions [2–21] on tribological behavior of PA composites have been analyzed. Several papers have reported that the addition of fiber reinforcement, such as glass and carbon fiber, helps in improving the tribological behavior of PA matrix, since some mechanical properties, such as strength and stiffness, are usually favored by the incorporation of fiber reinforcement [2–10, 14–16, 18]. Glass fiber has relatively low stiffness. However, its tensile strength is competitive with other fibers and its cost is much lower [2]. This combination of properties is likely

https://doi.org/10.1515/9783110352986-002

to ensure that glass fiber remains the most widely used reinforcement for high-volume commercial polymers applications. In the same sense, solid lubricants such as molybdenum disulfide (MoS_2) have been widely used as fillers on polymeric matrix to decrease the friction coefficient and wear rate of the polymers [4, 7, 11–13]. Use of these additives has also been explored in order to avoid the constant problems of environmental contamination. However, the improvement of tribological performance of PA composites is not always verified [3, 5, 10, 19–21]. In the case of MoS_2, some studies have argued that there is no advantage in adding MoS_2 to the PA, since it does not substantially reduce friction and causes an increase in wear [20]. It has even been argued that the addition of MoS_2 can cause increased wear and a decrease in friction, depending on the test conditions, and its behavior is reflected in the stability and thickness of the generated film [5]. In fact, friction and wear are not intrinsic properties but are strongly dependent on composition materials and the tribological tested conditions. In this respect, the tribological applications of PA composites, with steel as a counterpart, are required more in dry operating conditions than in other wear situations. These materials are often used in nonlubricated bearings, seals, cams, among others [3]. Therefore, most tribological investigations have been carried out on PA composites under dry sliding conditions [22]. The tribological behaviors of materials are related to the transfer film on counterface surfaces. After consolidating a continuous polymer layer on the metal surface, the counterface roughness and adhesion mechanisms are modified, allowing better tribological performance of the system. However, identifying homogeneity in transfer film as well as the analysis of the worn surface of the tribological systems is a complex task, requiring high expertise. In general, analysis of phenomena occurring on the counterface/worn surfaces is performed subjectively [23], which sometimes fails to yield the desired results. The analysis of the transfer of the film can also be evaluated by a technique that involves stylus-based roughness measurements along profile lines through the counterface [24, 25]. Although this distribution of the peaks and valleys detected during probe exploration at sample length may not always be related to the thickness of the transfer film, since in regions where there was no material adhesion the roughness of the counterface is registered. Alternatively, it is possible to draw straight lines directly over the microscopic images of the counterface microscopic images in order to detect the *distribution* of pixel intensity and associate them with the layer of material that has been adhered to the counterface. This procedure must be combined with a statistical criterion analysis of the variation of the intensity of pixels in the directions of greater transfer variability of the transfer film. The application of this criterion allows the exploitation of available information on counterface images after the tribo-tests with each material and, in this way, relates, in a comparative way, the uniformity of the transfer film with the respective tribological performances of the tested materials. In this study, the effects of the addition of glass fiber reinforcement and that of MoS_2 on the tribological behavior of PA66 were compared. The experiments were carried out on a pin-on-disk tribo-meter, in dry sliding conditions, and the material of counterface disk was a steel Ck45K. A set of test conditions was selected based on a pv limiting value (empirical parameter above which the wear rate increases rapidly) for the PA66 sliding against a steel counterface. The intent of this experiment is to approximate the test conditions needed to meet industrial demand. Comparison of the tribological behavior of materials was carried out by analyzing the uniformity of transfer film on the counterface, based on a new proposed approach of a statistical criterion analysis of the *distribution* of pixel intensity. Thus, the two specific goals of this work were: (1) to propose an expeditious methodology for the characterization and analysis of the uniformity of the transfer film on the counterface following the tribological tests, based on the extraction of the profiles of pixel intensities and statistical criterion of analysis; and (2) to study particularly the effect of the addition of MoS_2, since the results in the literature are relatively scarce and somewhat contradictory.

2.2 Materials and methods

The present study was conducted using the following commercial materials: PA66, PA66 + MoS_2 and PA66GF30. Table 2.1 presents the physical–chemical properties

Table 2.1: Physical–chemical properties of materials.

Properties	Standards	PA66	PA66 + MoS_2	PA66GF30
Density, g/cm³	ISO1183	1.14	1.15	1.29
Water absorption: Saturation at 23 °C % Saturation at 23 °C/50% RH	ISO 62	8 2.4	7.8 2.3	5.5 1.7
Melting temperature (DSC, 10 °C/min) °C		260	260	260
Thermal conductivity at 23 °C W/(K m)		0.2	0.2	0.3
Linear coefficients of thermal expansion (between 23° and 60 °C) m/(m K)		80×10^{-6}	80×10^{-6}	50×10^{-6}
Linear coefficients of thermal expansion (between 23° and 100 °C) m/(m K)		95×10^{-6}	90×10^{-6}	60×10^{-6}
Maximum allowed temperature of deflection under load at 23 °C 50% RH: (°C)	ISO 75	85	85	150
Maximum allowable temperature in air (short periods) °C (continuously: for 5,000/20,000 h) °C		180 95/80	180 110/120	240 80/95
Hardness (Rockwell test)	ISO 2039-2	M88	M76	M88

Note: http://www.lanema.pt (accessed January 10, 2018).

of the materials. Friction and wear behavior of the polymers were carried out on a Pin-on-Disk Tribo-meter. Cylindrical pin-shaped samples of PA66 and PA66 + MoS_2 (diameter of 15 mm and height of 10 mm) and PA66GF30 (diameter of 15 mm and height of 10 mm) were allowed to slide against a rotating disk (diameter of 76 mm and height of 8 mm). The counterface material was steel DIN-Ck45K-with 0.45% C, 0.25 Si and 0.65 Mn; the average surface hardness was 220HB; and surface roughness, $R_a \approx 0.44$ μm. Tribo-tests used the same contact pressure (1.46 MPa), sliding velocity (0.48 m/s) and sliding distance (7,500 m). The other detailed conditions are listed in Table 2.2. Limiting pv value chosen for this experiment was of 0.7 MPa m/s [26]. The morphological aspects on counterface and worn polymeric surfaces after the tests were examined by optical microscopy.

Table 2.2: Tribo-test conditions used in the studies.

Pin		Test conditions		
Material	**D (mm)**	**M (g)**	**T (°C)**	**RH %**
PA66	12.20	1.33	22	63
PA66GF30	15.90	2.55	24	60
PA66 + MoS_2	12.78	1.49	22	63

2.3 Development of statistical criterion for distribution of pixel intensity on counterface surfaces

Optical microscopic images of the counterface surfaces after the tribological tests were taken for three materials, under the conditions mentioned in Table 2.2. From these images, the distribution of pixel intensity was obtained and extracted along a set of equally spaced straight lines perpendicular to the sliding direction. As an example, Figure 2.1 shows a micrograph of the counterface surface after the tribological test with PA66, together with one of the data extraction lines used and the corresponding result (the *X*-axis represents distance along the line and the *Y*-axis represents the pixels intensity). This direction perpendicular to the sliding was selected because it makes possible to scan objectively the variations in uniformity in film formation. The decision to inspect several lines in the same direction on the image aimed to detect possible heterogeneity in the formation of the film. The pixel intensity extracted along the lines was analyzed using the distribution percentiles, illustrated as box plots and as scatter plots of percentile 25% versus percentile 75%. The box plots show the interquartile range (the box containing 50% of the data, between the percentiles 25% and 75%), the median and mean (line and dot inside the box) and whiskers corresponding to the percentiles 5% and 95%. The relative positioning of the different

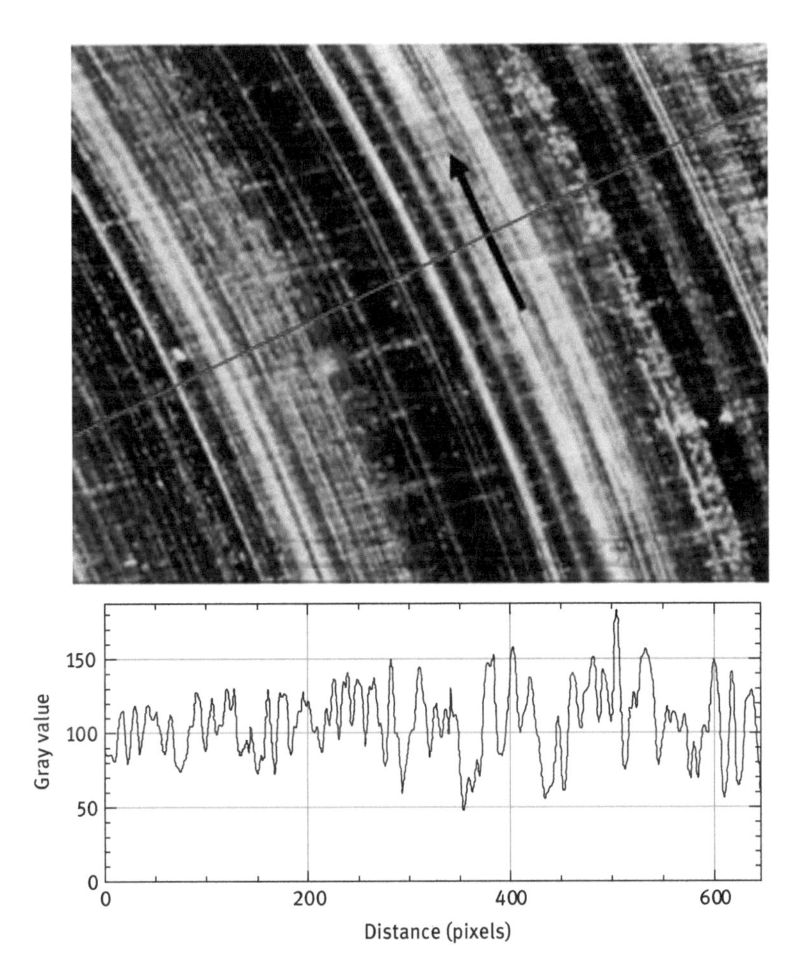

Figure 2.1: Image of counterface surface obtained by optical microscopy after the tribological test for PA66 and graph of pixel intensity along an illustrative line perpendicular to the sliding.

percentiles indicates the degree of dispersion and skewness in the pixel intensity or gray-level distribution. The scatter plots provide a picture of the range of data variability and facilitate the comparison among the different samples.

2.4 Results and discussion

Figures 2.2 and 2.3 show, respectively, the evolution of the friction coefficient as a function of the sliding distance and the evolution of temperature as a function of the sliding distance, for the three materials (PA66, PA66 + MoS_2 and PA66GF30) under the conditions described in Table 2.1.

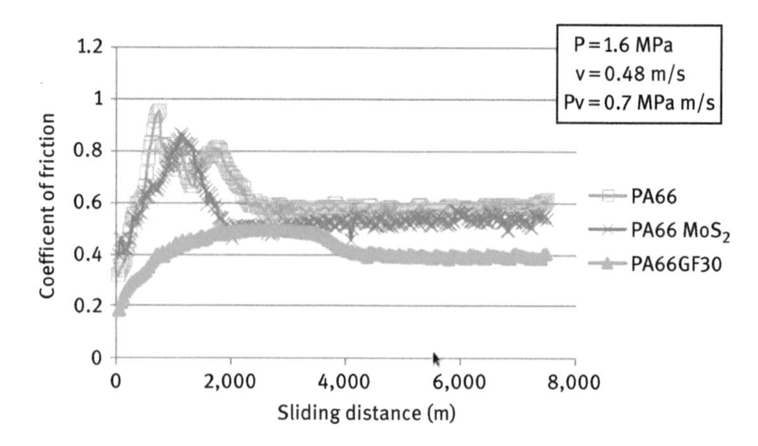

Figure 2.2: Evolution of the friction coefficient of PA66, PA66 + MoS_2 and PA66GF30 as a function of sliding distance.

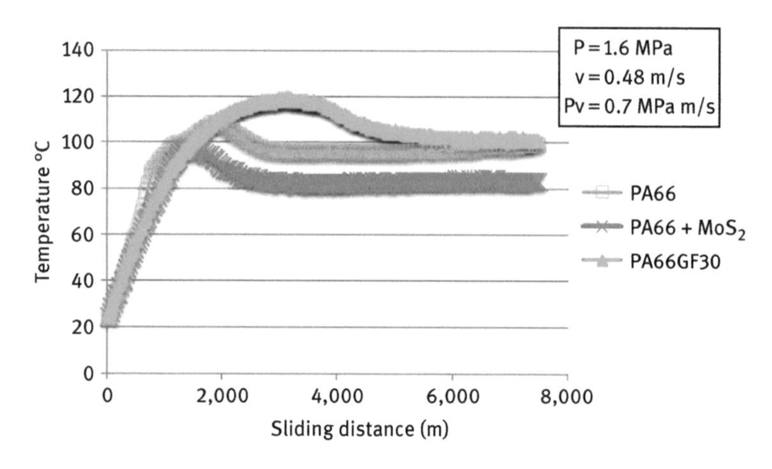

Figure 2.3: Temperature evolution of PA66, PA66 + MoS_2 and PA66GF30, as a function of sliding distance.

In general, the friction coefficients of all specimens exhibit an increase during the initial phase of the experiment, followed by decreasing values during an intermediate period of operation and then entering a stable regime (Figure 2.2). The friction coefficient values for both samples, PA66 and PA66 + MoS_2, increase remarkable for values between 1,000 and 2,000 m sliding distance, followed by a sharp decrease. It is known that the increase in temperature and friction coefficient is interconnected, and their behavior throughout the experiment is identical (Figures 2.2 and 2.3). The temperature in the contact zone increases proportional to friction. In PAs both parameters have a strong increase in the initial state of wear and are a consequence of the increase in the contact area and the heating by friction [15, 27]. The PA66 matrix has a low glass transition point [28] and low thermal conductivity (Table 2.1). These two parameters associated with heat releasing from the friction rapidly lead to the melting temperature. When the polymer melts, its wear and friction behavior are drastically altered. The friction coefficient tends to decrease and the rate of wear tends to increase with increasing temperature [2]. On the other hand, the friction coefficient of Pa66GF30 presented a smoother evolution as a function of the sliding distance, in comparison with PA66 and PA66 + MoS_2. In addition, the PA66GF30 sample presented a lower friction coefficient than the other two samples under dry sliding conditions (Figure 2.2).

The analysis of counterface surfaces is fundamental to understand the tribological behavior of these three specimens. Figure 2.4 presents the sequence of optical microscope images of counterface surface after the sliding of each material (Pv = 0.7 MPa m/s). Figure 2.4(a) shows an image of the grinding disk by comparison.

In a visual inspection, it can be observed that although the formation of tribo-film was predominant on the counterface in the three tribological tests, the material deposition was quite different. In Figure 2.4(b), it can be seen that PA66 adhered to the steel and formed a film of very irregular material. In the case of PA66 + MoS_2 (Figure 2.4(c)), it was observed that there was adhesion of material which provided the formation of the film; however, this film presents some flaws. Regarding the sample PA66GF30, it apparently exhibited a more uniform transfer film (Figure 2.4(d)).

The investigation of the uniformity of the film transfer on the surface was done using the distribution percentiles of pixel intensity. Figure 2.5 shows the comparison of box plot of variation in pixel intensity on counterface for specimens in three different blocks: PA66, PA + MoS_2 and PA66GF30. Figure 2.6 shows the corresponding scatter plot for the percentiles 25% versus 75%. The interquartile range (IQR) shown as a rectangular box in the box plots measures the statistical dispersion of pixel intensity comprising 50% of the data. Thus, the smaller the IQR, the smaller range of variability and a priori the more homogeneous distribution of pixel intensities and the more uniformly distributed the transferred film will be. Figure 2.5 shows clear differences in the homogeneity of the pixel intensity for the three samples. It is also noticed that the variation along the micrograph within the same sample is smaller than the variation among samples, which means that there is no significant heterogeneities in the slide direction. Sample PA66 exhibits a very larger IQR compared to the two other samples: PA66 + MoS_2 and PA66GF30. The irregularity of the transfer film by sample PA66 can also be evaluated by larger size of whisker, which provides information on the dispersion of data (Figure 2.5). It could explain the initial unstable evolution of the friction coefficient of the PA66 sample (Figure 2.3). The film formed by PA66 + MoS_2 has better stability when compared to the film formed by PA66. It presents a considerably smaller IQR, which may be associated with the fact of its friction coefficient to be slightly lower than PA66. At the opposite side is the

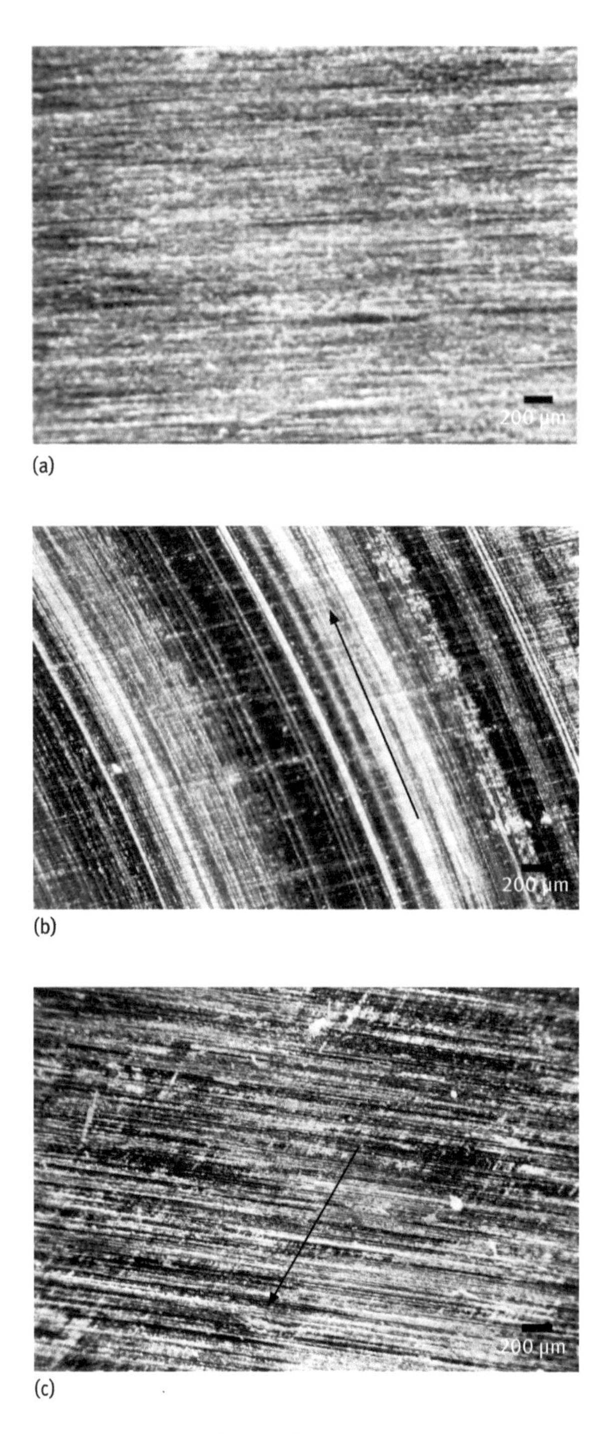

Figure 2.4: Images of counterface surfaces after tribological tests. The arrows represent the sliding direction: (a) Grinding disk surface; (b) PA66; (c) PA66+ MoS_2; and (d) (d) PA66GF30.

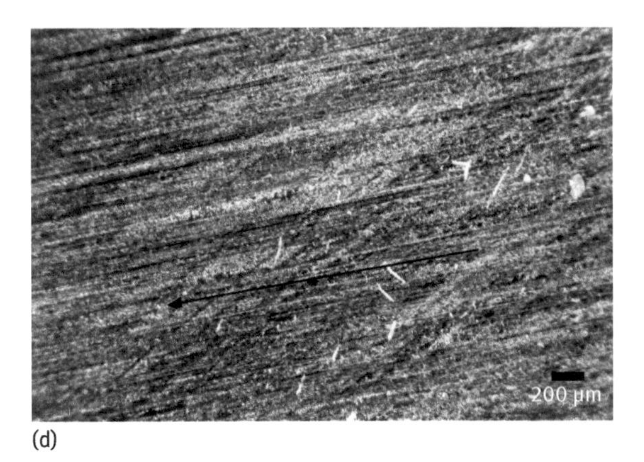

(d)

Figure 2.4: (continued)

PA66GF30, which presents the better friction coefficient behavior (Figure 2.3). This arises from the more continuous dispersion of the transfer film (Figures 2.4, 2.5). It is known indeed that the friction coefficient can be decreased and the wear resistance can be enhanced by adding glass fiber in PAs [29]. Furthermore, glass fiber improves the load-carrying capacity and the thermal conductivity, which leads to a positive effect of lowering wear rate and provides good performance to PA66 [27]. This can be observed in Figure 2.7, which presents the wear coefficient, W, for PA66, PA66 + MoS_2 and PA66GF30 samples.

The fact that the percentile differences within the same sample are smaller than the differences between samples is clarified in Figure 2.6, showing the scatter plot of percentile 25% versus percentile 75% for all the transects analyzed (i.e., all the box plots of Figure 2.5). Using the points of the same sample, a convex hull polygon was created around them and displayed in Figure 2.6. No overlapping of the three polygons is observed, indicating the three samples are clearly separated in the scatter plot of percentiles. Finally, the observation of Figure 2.4(d) suggests that the wear mechanisms involved are tribo-film, adhesion abrasion. In fact, during the sliding the fibers are exposed repeatedly to flexion, which causes fibers that are transformed into wear particles. Figure 2.8 displays an optical micrograph of PA66GF30 worn surface with fiber debris.

Table 2.3 shows the compilation of the experimental tribological results of the friction coefficient, μ, the wear coefficient, W, final temperature, T_f, and the maximum temperature, T_m, after a sliding distance of 7,500 m, recorded during the test for the three materials. Maximum temperature reached by PA66 and PA66 + MoS_2 can be seen above the deformation temperature limit under load suggested by the manufacturer (85 °C) (Table 2.1), but in PA66GF30, they remain below the deformation temperature under suggested load by the manufacturer (150 °C) (Table 2.1). It

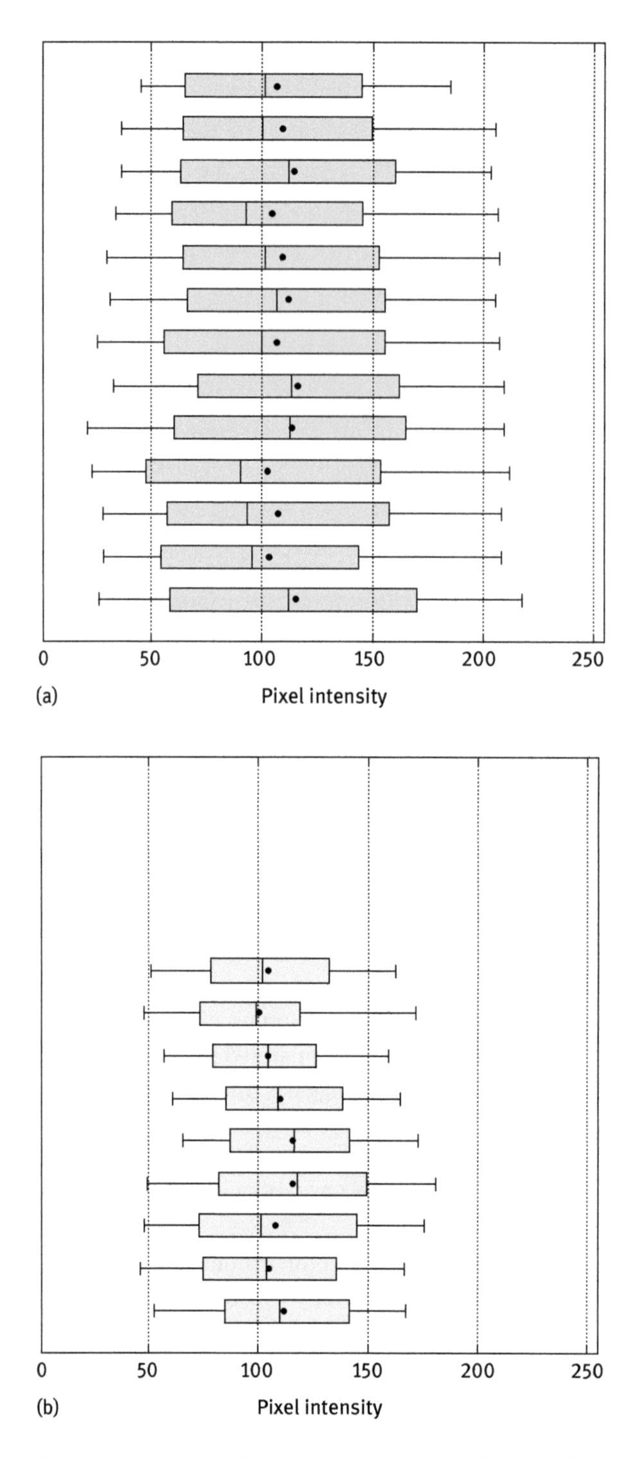

Figure 2.5: Pixel intensity box plot for the samples (a) PA66, (b) PA + MoS_2 and (c) PA66GF30.

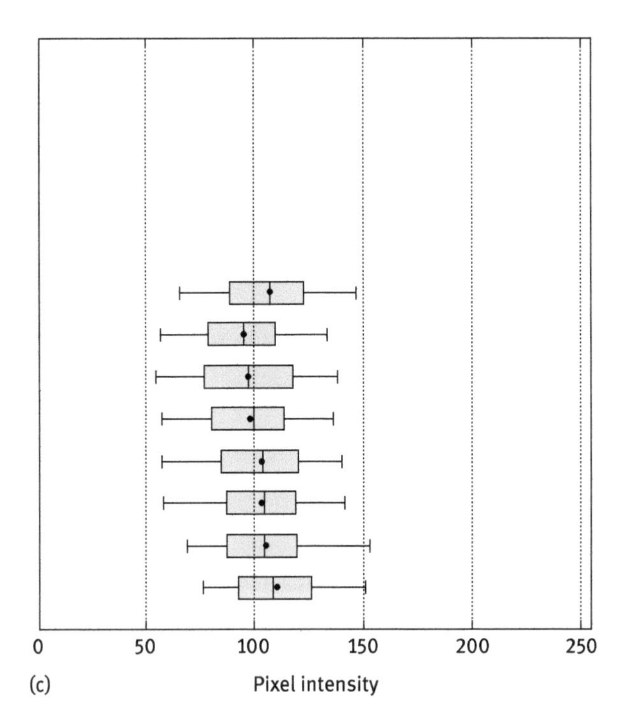

Figure 2.5: (continued)

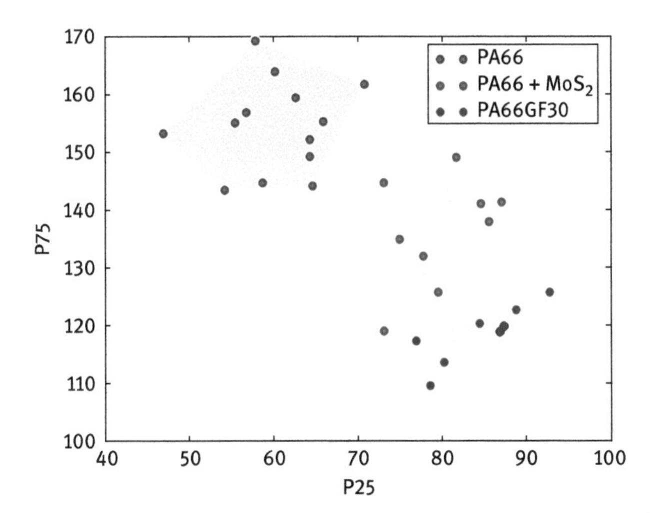

Figure 2.6: Percentile 75% (P75) versus percentile 25% (P25) for the box plots illustrated in Figure 2.5. The shadow areas show the convex hull of the three sets of points corresponding to the three samples analyzed.

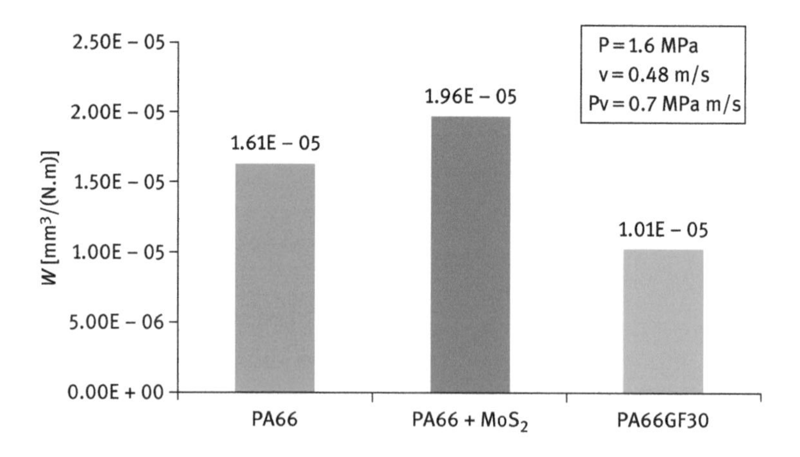

Figure 2.7: Wear coefficient, *W*, of PA66, PA66 + MoS_2 and PA66GF30.

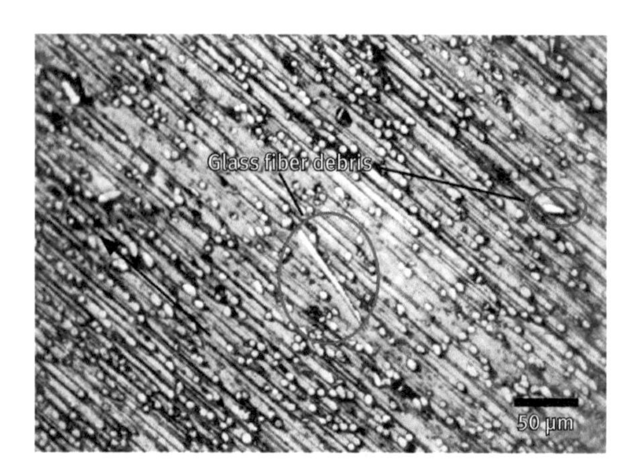

Figure 2.8: PA66GF30 worn surface with exposed wear debris. The arrows represent the sliding direction.

Table 2.3: Experimental tribological results.

Material	*W* (mm³/(N m))	μ	T_f (°C)	T_m (°C)
PA66	1.61×10^{-5}	0.60	99	108
PA66 + MoS_2	1.96×10^{-5}	0.54	83	97
PA66GF30	1.01×10^{-5}	0.39	101	119

should be noted that these temperatures are measured on the face of the disk by a pyrometer and therefore are lower than the contact temperature; hence, this parameter is only a reference.
The friction coefficient of PA66 + MoS_2 decreased slightly in relation to PA66, while its wear resistance performance worsened; on the other hand, its temperature in the contact zone remained lower throughout the tribological test (Figure 2.3). These results suggest that the addition of this self-lubricant additive to *PA66* decreases significantly the temperature in the contact zone and slightly the coefficient of friction at the cost of a large increase in the wear rate of the material, for the present test conditions. In this way, the use of PA66 + MoS_2 is recommended in situations where temperature is a determining factor.

2.5 Conclusions

In this work we proposed a new approach for the analysis of the uniformity of the transfer film on the counterface in a tribological test, using a statistical criterion based on the analysis of the *distribution* of pixel intensity. The relationship between the transfer film on the counterface and the tribological behavior of PA66, PA66 + MoS_2 and PA66GF30 was analyzed. Based on the experimental results presented and the statistical analysis performed the following conclusions can be drawn:

- The glass fiber reinforcement leads to positive effects on the tribological behavior of PA66. It was verified that friction coefficient and wear rate of the composites decreased. Glass fiber improves the load-carrying capacity and the thermal conductivity, which provides good performance to PA66.
- The addition of MoS_2 to PA66 decreases significantly the temperature in the contact zone and slightly the coefficient of friction at the cost of a large increase in the wear rate of the material, for the present test conditions. In this way, the use of PA66 + MoS_2 may be advantageous in situations where temperature is a determining factor.
- The dominant wear mechanism for all tribological tests was the tribo-film formation, but the homogeneity of transfer film on counterface was quite different for each specimen. The investigation of the uniformity of the film transfer on the counterface is associated with the tribological behavior of the materials and can be inspected quantitatively using the distribution percentiles of pixel intensity.
- The P25 versus P75 scatter plots show the points aligned from low P25/high P75 for PA66 to high P25/low P75 for PA66GF30, i.e., the IQR decreases and the data becomes more concentrated toward higher pixel intensities.
- The statistical analysis employed proved to be useful and robust for the evaluation of the tribological behavior of PA66 composites at the described conditions. The relative positioning of the different percentiles indicate the degree of dispersion and skewness in the pixel intensity or gray-level distribution. The scatter

plots provide a picture of the range of data variability and facilitate the comparison among the different samples.
- It is certainly necessary to test the relationship between the results obtained by the statistical analysis used and the friction and wear performance of many other tribological systems to further validate the approach. However, the good results obtained here give us confidence on the robustness of the method and we believe that it will find widespread use.

Acknowledgments
Ana Horovistiz acknowledges the TEMA (Centre for Mechanical Technology and Automation), Department of Mechanical Engineering for funding the scholarship UID/EMS/00481/2013.

References

[1] Aldousiri B, Shalwan A, Chin CW. A review on tribological behaviour of polymeric composites and future reinforcements. Adv Mater Sci Eng 2013;DOI:10.1155/2013/645923.
[2] Kim JW, Jang H, Woo Kim J. Friction and wear of monolithic and glass-fiber reinforced PA66 in humid conditions. Wear 2014;309:82–8. DOI:10.1016/j.wear.2013.11.007.
[3] Kukureka SN, Hooke CJ, Rao M, Liao P, Chen YK. Effect of fibre reinforcement on the friction and wear of polyamide 66 under dry rolling-sliding contact. Tribol Int 1999;32:107–16. DOI:10.1016/S0301-679X(99)00017-1.
[4] Xing Y, Zhang G, Ma K, Chen T, Zhao X. Study on the friction and wear behaviors of modified PA66 composites. Polymer Plast Tech Eng 2009;48:633–8. DOI:10.1080/03602550902824481.
[5] Wang J, Gu M, Songhao B, Ge S. Investigation of the influence of MoS_2 filler on the tribological properties of carbon fiber reinforced nylon 1010 composites. Wear 2003;255:774–9. DOI:10.1016/S0043-1648(03)00268-0.
[6] Samyn P, Schoukens G. Thermochemical sliding interactions of short carbon fiber polyimide composites at high pv-conditions. Mater Chem Phys 2009;115:185–95. DOI:10.1016/j.matchemphys.2008.11.029.
[7] Tang G, Huang W, Chang D, Nie W, Mi W, Yan W. The friction and wear of aramid fiber-reinforced polyamide 6 composites filled with nano-MoS_2. Polym Plast Technol Eng 2011;50:1537–40. DOI:10.1080/03602559.2011.603779.
[8] Rubio J, Silva L, Leite W, Panzera T, Filho S, Davim J. Investigations on the drilling process of unreinforced and reinforced polyamides using Taguchi method. Composites: Part B 2013;55:338–44. DOI.org/10.1016/j.compositesb.2013.06.042.
[9] Njuguna J, Mouti Z, Westwood K. Toughening mechanisms for glass fiber-reinforced polyamide composites. Elsevier Ltd., 2015. DOI:http://dx.doi.org/10.1016/B978-1-78242-279-2.00008-1.
[10] You Y-L, Li D-X, Deng X, Li W-J, Xie Y. Effect of solid lubricants on tribological behavior of glass fiber reinforced polyamide 6. Polym Compos 2013;34:1783–93. DOI:10.1002/pc.22582.
[11] Liu B, Pei X, Wang Q, Sun X, Wang T. Applied surface science effects of proton and electron irradiation on the structural and tribological properties of MoS_2/polyimide. Appl Surf Sci 2011;258:1097–102. DOI:10.1016/j.apsusc.2011.09.041.

[12] Rodriguez V, Sukumaran J, Schlarb AK, De Baets P. Influence of solid lubricants on tribological properties of polyetheretherketone (PEEK). Tribol Int 2016;103:45–57. DOI:10.1016/j.triboint.2016.06.037.
[13] Eleiche AM, Mokhtar MO, Kamel GM. Developing a new polyamide composite to solve tribological problems associated with rotating bands. Procedia Eng 2013;68:231–7. DOI:10.1016/j.proeng.2013.12.173.
[14] Li J, Sheng XH. The effect of PA6 Content on the mechanical and tribological properties of PA6 reinforced PTFE composites. J Mater Eng Perform 2010;19:342–6. DOI:10.1007/s11665-009-9485-8.
[15] Quaglini V, Dubini P, Ferroni D, Poggi C. Influence of counterface roughness on friction properties of engineering plastics for bearing applications. Mater Des 2009;30:1650–8. DOI:10.1016/j.matdes.2008.07.025.
[16] Bijwe J, Naidu V, Bhatnagar N, Fahim M. Optimum concentration of reinforcement and solid lubricant in polyamide 12 composites for best tribo-performance in two wear modes. Tribol Lett 2006;21:57–64. DOI:10.1007/s11249-005-9010-7.
[17] Bermúdez MD, Carrión-Vilches FJ, Martínez-Mateo I, Martínez-Nicolás G. Comparative study of the tribological properties of polyamide 6 filled with molybdenum disulfide and liquid crystalline additives. J Appl Polymer Sci 2001;81:2426–32. DOI:10.1002/app.1683
[18] Davim J, Silva L, Festas A, Abrão AM. Machinability study on precision turning of PA66 polyamide with and without glass fiber reinforcing. Mater Des 2009;30:228–34. DOI.org/10.1016/j.matdes.2008.05.003.
[19] You Y.L, Li DX, Si GJ, Deng X. Investigation of the influence of solid lubricants on the tribological properties of polyamide 6 nanocomposite. Wear 2014;311:57–64. DOI:10.1016/j.wear.2013.12.018.
[20] Zhang SW. State-of-the-art of polymer tribology. Tribol Int 1998;31:49–60. DOI:10.1016/S0301-679X(98)00007-3.
[21] Stachowiak GW, Batchelor AW, Stachowiak GW, Batchelor AW. 16 – Wear of non-metallic materials. Eng Tribol 2006;651–704. DOI:10.1016/B978-075067836-0/50017-1.
[22] Pogačnik A, Kupec A, Kalin M. Tribological properties of polyamide (PA6) in self-mated contacts and against steel as a stationary and moving body. Wear 2017;378–9:17–26. DOI:10.1016/j.wear.2017.01.118
[23] Soleimani S, Sukumaran J, Kumcu A, De Baets P, Philips W. Quantifying abrasion and micro-pits in polymer wear using image processing techniques. Wear 2014;319:123–37. doi.org/10.1016/j.wear.2014.07.018.
[24] Deleanu L, Cantaragiu A, Podaru G, Georgescu C. Evaluation of the spread range of 3D parameters for coated surfaces. Tribol Ind 2011;33:72–8.
[25] Deleanu L, Ciortan S., Andrei G, Maftei L. Study on the profilomtry of composites with pa matrix and micro glass spheres after dry sliding. In Vencl A, Marinković A. editors. 11th International Conference on Tribology – SERBIATRIB'09. Belgrade, Serbia, 2009.
[26] Van De Velde F, De Baets P. The friction and wear behaviour of polyamide 6 sliding against steel at low velocity under very high contact pressures. Wear 1997;209:106–14. doi:org/10.1016/S0043-1648 (96)07500-X
[27] Abdelbary A. Wear of polymers and composites. Oxford: Elsevier, 2014. DOI:10.1533/9781782421788.1.
[28] Wypych G. PA-6, 6 polyamide-6, 6. G. Wypych, (Ed.), In Handbook of polymers. Oxford: Elsevier, 2012:215–20. DOI: 10.1016/B978-1-895198-47-8.50070-9
[29] Demirci MT, Düzcükoğlu H. Wear behaviors of polytetrafluoroethylene and glass fiber reinforced Polyamide 66 journal bearings. Mater Des 2014;57:560–7. DOI:10.1016/j.matdes.2014.01.013.

T.S. Kiran, S. Basavarajappa, M. Prasanna Kumar, B.M. Viswanatha, and J. Paulo Davim

3 A review on dry sliding wear behaviour of metal matrix composites

3.1 Introduction

All moving parts experience wear, a phenomenon that affects the performance of any component under direct contact. Not all components can be lubricated regularly, leading to friction and eventually the size and shape of the component reaches a point where the intended operation is impossible. In order to get the mechanism working, the worn out components should be replaced leading to increase in operating cost and time. Generally hybrid composites are preferred over metals, alloys and even composites nowadays as the hybrid composites comprises of two or more reinforcements. The particulate reinforced aluminium metal matrix composites (MMCs) offer a wide range of properties that are suitable for numerous engineering applications. To have an extensive knowledge on the work that has been carried out by earlier researchers on processing techniques, mechanical properties and wear behaviour of MMCs are reviewed. Several researchers are of the view that, change in mechanical properties influences the wear behaviour significantly. The improvement in mechanical and tribological properties needs an intense research as slight variations in composition have shown superior performance. In contrast, minor changes in composition have also reduced the desired properties. Several materials improve their mechanical and tribological properties without changing the chemical combination and attain the desired properties vis-à-vis changing the microstructure by heat treatment/heat treated (HT) or ageing. In view of this, a comprehensive evaluation of variation in properties due to modifications made in metals, alloys or composites are studied.

3.2 Mechanical properties

For developing structural materials possessing the combination of physical and mechanical properties, particulate reinforced MMCs provide significant opportunities that are not available with monolithic alloys. A wide range of matrices and reinforcements permits to develop MMCs with low density and high modulus with improved thermal conductivity. The improved mechanical properties of particulate reinforced MMCs like high tensile strength, fatigue and creep resistance make them attractive, but the reduction in ductility and fracture behaviour hinders the use of MMCs. The following section reviews the mechanical properties of MMCs.

https://doi.org/10.1515/9783110352986-003

3.2.1 Hardness

The processed MMCs are desired to attain higher hardness than the alloy. The reinforcements play an important role in enhancing the hardness. The common reinforcements that are used are silicon carbide particles (SiCp), alumina (Al_2O_3), titanium carbide (TiC), garnet, boron carbide (B_4C), graphite (Gr), Molybdenum disulphide (MoS_2) and so on. The inclusion of SiCp alone with various matrices and with various size, shape and weight percentage was carried out by several researchers [1–24]. The conclusion drawn was, as the weight percentage of SiCp increases, the hardness increased as it is a hard reinforcement. The hard SiCp acts as barrier to the movement of dislocations in the alloy. The inclusion of Gr into alloy considerably reduced the hardness as it is a soft dispersoid [25–30].

The combination of hard SiCp and soft Gr particles into the alloy, in varying weight percentages [26, 31–36], showed an improvement in hardness. A contrast result was observed in hybrid composites [37, 38] where the hardness was reduced as Gr percentage was increased. A reduction in hardness was noticed after HT for alloy [39, 40, 41–46], composite [6, 47] and ageing [39, 47]. On the contrary, hardness improved after HT [5] for aluminium alloy and ZA-27/Gr reinforced composite after ageing [25].

3.2.2 Tensile properties

The evaluation of tensile strength of hard and soft reinforcement has been carried out by several researchers. The values extracted from tensile test reveals the strength of the material processed and facilitate to select a material with optimum strength. Reinforcing of particles (SiCp, Al_2O_3, Zircon, TiC, TiO_2, Gr and so on) alone in aluminium alloys enhance ultimate tensile strength (UTS) and reduces ductility [6, 48, 49]. The reason for improvement in UTS of the composites is the inclusions that act as barriers to the dislocations in the microstructure resulting in decrease in inter-particulate distance and leading to dislocation pile-up. The tensile properties vary with varying strain rates and testing temperatures [50]. At slow strain rate, the tensile properties deteriorated while at higher strain rate the tensile properties showed improved results. At higher temperature, UTS reduced while ductility improved. The HT and aged specimen showed reduced UTS [42] and improved ductility. At low strain rates, the microporosity and inclusions actively participate than at higher strain rates.

3.3 Wear behaviour

The designed component should perform to its fullest capacity in a machine and the life of such component design is a critical issue for a designer. The selection of material for an application varies depending on the specific strength and modulus, cost, density and atmospheric condition in which the material operates. Metals and alloys are the major materials used to manufacture components like engine components, gear drives and so on in industries such as automotive and aerospace. The conditions under which these components work may be either in lubricated or sometimes even in dry conditions. But when the condition is dry, it results in more wear and tear of the components that may eventually lead to catastrophic failure. Wear of an element is the major problem that needs to be addressed to improve the life of a product. The incapability of metals and alloys to withstand various loads applied led to the evolution of composite materials. The addition of reinforcements has drastically improved the tribological as well as mechanical properties. This section reviews the wear behaviour of alloys, composites and hybrid composites for both as-cast and HT specimens at various applied loads, sliding speeds and sliding distances.

3.3.1 As-cast

As-cast material is a condition in which the material under consideration has not undergone any heat treatment other than basic facing, turning and finishing operations. The cast material is brought to a standard dimension that undergoes a set of tests. This section reviews the wear behaviour of alloys and composite specimens under as-cast condition.

3.3.1.1 Wear behaviour of SiCp reinforced MMCs

The wear behaviour of an alloy and composite is studied to know the effect of applied load, sliding speed and sliding distance on the specimen. The type, size, shape and weight percentage of reinforcement added are the influential factors on wear behaviour. The advantages and disadvantages are discussed and arrived for possible conclusions. An improvement in the wear behaviour was discussed meticulously with the addition of SiCp alone by several researchers.

Several investigators [1, 7–24] evaluated the wear behaviour of aluminium alloy and compared it with SiCp reinforced composite. The results revealed an improvement in the wear resistance as the percentage of SiCp increased. An increase in the wear behaviour of the specimen was observed as the load and speed were increased. The size of the reinforcement considered had greater influence on the wear resistance. Higher wear resistance was observed for large particles as they participate actively in the wear by initially fragmenting and as the size gets eroded, they detach from the alloy. The smaller particles are detached easily as the contact area is low [7, 13, 17, 18, 21].

Sharma et al. [1] studied the dry sliding wear behaviour of zinc alloy (ZA-27) reinforced with SiCp at varying load (3, 4, 5 and 6 N) and sliding speed (1.25, 1.56 and 1.87

m/s). Liquid metallurgy technique was employed to fabricate the specimen with particle size (20–30 μm) and with varying wt.% (1, 3 and 5). The wear resistance improved with an increase in the wt.% of reinforcement and sliding speed, but lowered as applied load was increased. Abrasion wear behaviour was observed at lower load, while particle cracking induced delamination wear behaviour dominated at higher loads. The reason is the initiation and propagation of surface cracks that deteriorates wear resistance at higher loads. The hardness and wear behaviour of the composite specimen enhanced substantially due to the addition of reinforcement.

Prasad et al. [9, 12, 15] studied the sliding wear behaviour of zinc matrix (ZA-37.5) reinforced with SiCp at varying load (5–140 N) and constant sliding speed (2.68 m/s). Stir casting method was considered to fabricate the specimen with particle size (50–100 μm) and 10 wt.% of SiCp. Composite specimen exhibited less rise in temperature, lower friction co-efficient and excellent wear behaviour than the base alloy at all applied load. Deep grooves and microcracks were found on the surface of alloy, while shallow grooves were observed on composite specimen. Aluminium rich (α) acts as load bearing phase while zinc rich (η phase) acts as solid lubrication [9]. As SiCp is harder and thermally stable than the base alloy, wear resistance of the composite is better than the base alloy.

Kumar and Balasubramanian [7, 17] evaluated the sliding wear behaviour of AA7075 aluminum matrix embedded with varying size and wt.% of SiCp. Larger particles (149 μm) revealed superior wear resistance in comparison with the smaller particles (44 μm). At lower speed (0.3 m/s) and load (52 N), higher wear resistance was observed than at higher speed (1.5 m/s) and loading (152 N) condition. Small particles were more suitable for low speed applications as the particles adhere with the alloy leading to superior wear resistance [17]. Vieira et al. [14] evaluated the sliding wear behaviour of pure aluminum matrix reinforced with SiCp (37 μm). Composite specimen was fabricated by stir casting process followed by centrifugal casting, so that higher hardness is desirable at the outer surface as the particles tend to move towards the surface due to centrifugal action. The wear test results revealed the formation of mechanically mixed layer (MML) that constituted iron, iron oxide, aluminum oxide and intermetallic compounds of Al-Fe and Al-Fe-O.

Lu et al. [16] explained in brief the formation of MML due to material transfer in pure aluminium reinforced with SiCp. Composite specimen was fabricated by vacuum pressure infiltration technique. The debris emerging from wear process is entrapped between the specimen and the rotating disc. Due to large difference in the hardness, the debris was pushed into the grooves of specimen and was flattened forming MML. Wilson and Alpas [11, 22] evaluated the sliding wear behaviour of Al6061/Al_2O_3 and A356/SiCp with 20 wt.% of reinforcement. A356/SiCp exhibited superior wear resistance than Al6061/Al_2O_3 due to higher hardness and fracture resistance of SiCp. Al_2O_3 particles were prone to fracture than SiCp. Ultra mild wear was observed for composites, while severe and mild wear was experienced for alloy. Temperature rise was the influential factor in the wear behavior.

Venkataraman and Sundararajan [24] gave a detailed explanation on the formation of MML on the surface of the composite specimen. The comparison for HT, peak-aged composite (10 and 40 wt.% SiCp) specimen was carried out on pin-on-disc arrangement in dry sliding condition. The sliding speed (1 m/s) and sliding distance (4,000 m) were constant while the loads were varied (50–280 N). A thorough investigation was made on friction co-efficient, thickness of MML, formation and rupture of the MML that delineate mild and severe wear. The hardness of MML was very high than the bulk hardness resulting in improved wear resistance. The substantial conclusion drawn from the experimental work was the formation of ultra thin, steady and hard MML that provides superior wear resistance.

3.3.1.2 Wear behaviour of Gr reinforced MMCs

The instances of working under dry or unlubricated conditions are increasing. The presence of a solid lubricant under these conditions is the need of the hour. The soft solid lubricants like Gr or molybdenum-disulphide (MoS_2) are reinforced with the alloy to overcome the hindrance.

Akhlaghi and Zare-Bidaki [28] evaluated the dry sliding wear behaviour of graphite (5–20 wt.%) reinforced with A2024 alloy. The outcome of the investigation was that the porosity, hardness, bending strength and co-efficient of friction reduced. Wear test was carried out on pin-on-disc apparatus with an applied load of 50 N, sliding speed of 0.5 m/s and sliding distance of 1,000 m. The wear rate measured for 5 wt.% of Gr content was superior compared to the alloy and other composition (10, 15 and 20 wt.%) of composites. At higher wt.% of Gr particles (10, 15 and 20), the formation of cracks and deterioration of mechanical properties resulted in increased delamination. Seah et al. [29] evaluated the sliding wear behaviour of ZA-27/Gr particles (1, 3 and 5 wt.%) at various applied load (30, 40 and 50 N), sliding speed (1.25, 1.63 and 2 m/s) with duration of 15 min. The addition of Gr particles improved the wear resistance at the cost of hardness. A significant improvement in wear resistance was observed for the first 1% of reinforcement, while there was less benefit on further addition (3 and 5 wt.%) of Gr particles. As load was increased, wear resistance reduced, as the smeared Gr layer was unable to withstand the load. Wear resistance increased for higher sliding speed, as the time taken for the Gr particles to smear from the specimen was less.

Babic et al. [30] reinforced 2 wt.% of Gr particles to evaluate the sliding wear behaviour of ZA-27 matrix and composite. During the initial period of sliding, wear was intensive as it was observed in wear loss and co-efficient of friction. This was due to the minimum contact area at the initial stages (running-in process) that further improved once the steady state was achieved. As the applied load (10, 20, 30 and 50 N) and sliding speed (0.26, 0.5 and 1.0 m/s) increased, the wear rate also increased for alloy and composite specimen. The wear rate and friction co-efficient of composites were lower compared with the alloy.

3.3.1.3 Wear behaviour of Hybrid MMCs (HMMCs)

Wear behaviour of reinforcing SiCp or Gr alone leads to limited application. The addition of SiCp alone increases wear resistance as it takes the maximum load but makes the machining difficult and the composites become brittle. Addition of Gr alone improves wear resistance by reducing friction between sliding surfaces, as it is a solid lubricant but reduces the mechanical strength. Thus the combination of both (SiCp and Gr) can be advantageous in retaining mechanical strength, improving machining as well as wear behaviour.

Suresha and Sridhara [26, 31, 32, 36] evaluated the dry sliding wear behaviour of hybrid composite reinforced with equal wt.% of reinforcements (SiCp and Gr) into LM25 alloy. The results of hybrid composites were compared with addition of SiCp [36] and Gr alone [26, 32] in LM25 alloy to establish the advantage of the former. Elaborate studies were made on wear [26, 32 and 36] and friction characteristics [31] of HMMCs. An increase in the sliding speed reduces wear, while an increase in load increases wear. The results discussed were with respect to the importance of retention and removal of MML respectively. As sliding distance was increased, the unstable MML was responsible for increase in wear. The optimum reinforcement was 7.5% (3.75 wt.% SiCp and 3.75 wt.% Gr each) for any value of applied load, sliding speed and sliding distance. The results were independent of the friction of disc surface due to the formation of MML.

Three wear regimes (ultra-mild, mild and severe wear) were explained in detail based on the formation of MML by Riahi and Alpas [33]. The two compositions of hybrid composites considered for the dry sliding wear behaviour were A356-10%SiCp-4%Gr and A356-5%Al_2O_3-3%Gr. A wide range of load (0.4–420 N) and sliding speed (0.2–3 m/s) were used for the wear test at a constant sliding distance (6,000 m). The mass-based wear loss was more at higher load and speed for both composite specimens. The removal of MML was responsible for the transition from mild to severe wear. At lower loads and speeds (mild wear), the thickness of MML was maximum and the hardness was eight times the bulk hardness. But at higher load (severe wear) the thickness of MML were reduced considerably and the surface hardness also dropped drastically. SiCp-based hybrid composites offered superior wear resistance than Al_2O_3 at higher applied load and sliding speed. Kumar and Dhiman [34] evaluated the sliding wear behaviour of A7075 and A7075-7%SiCp-3%Gr in dry condition at varying load (20–60 N), speeds (2–6 m/s) and sliding distances (2,000–4,000 m). Hybrid composites revealed superior wear behaviour than the base alloy. Load was the most influential factor that negatively affected the wear resistance significantly, followed by sliding speed and sliding distance.

The influence of addition of Gr particles (5 and 10 wt.%) on Al2024/5%SiCp composites was demonstrated by Ravindran et al. [35 and 37]. On pin-on-disc apparatus dry sliding wear test were carried out at an applied load (10 and 20 N), sliding speed (1 and 2 m/s) and sliding distance (1,000 and 3,000 m) to assess the wear loss and friction coefficient. Al2024/5%SiCp and Al2024/5%SiCp-10%Gr composites exhibited higher wear loss and friction coefficient than Al2024/5%SiCp-5%Gr reinforced HMMCs, revealing the importance of limited addition of Gr particles. Excessive addition (10 wt.%) of Gr particles reduces the hardness and fracture energy thereby reducing the strength of the specimen resulting in increased wear loss. Sliding distance and load were the most influential factors on wear loss and friction coefficient respectively, while Gr content was the least. The dry sliding tribological behaviour of self–lubricating, Al6061/10%SiCp-5Gr-Ni-coated hybrid composites were evaluated by Guo and Tsao [38]. The nickel-coated graphite particles were uniformly distributed compared with uncoated. At higher Gr content (8%), the size of wear debris were smaller as the specimen was exhibiting brittle behaviour and the amount of Gr particles released were more compared to 2% and 5% addition resulting in less weight loss and friction coefficient. The wear rate of composite specimen increases for 2 wt.% and 5 wt.% of Gr particles and then reduces at 8 wt.% addition as the effect of SiCp abrading was reduced by excessive release of Gr particles.

A study on sliding wear behaviour of Al2219, Al2219/SiCp and Al2219/SiCp-Gr specimens was carried out by Basavarajappa et al. [51–54]. In addition to the evaluation of dry sliding wear behaviour, subsurface deformation and the effect of variation of microhardness along the depth, perpendicular to the worn surface, was evaluated. The wear resistance of hybrid composite (Al2219/15SiCp-3Gr) was pronounced to be more superior to the alloy and other composites. The microhardness of hybrid composites stabilized at much lower depth than the composite specimen revealing an improvement in wear resistance due to the addition of Gr particles [51]. To evaluate the contribution of factors (load, sliding speed and sliding distance) on wear behaviour, a DOE (design of experiment) based Taguchi technique was considered [52 and 54]. Sliding distance was the influential factor for both (Al2219/SiCp and Al2219/SiCp-Gr) specimen, followed by load and sliding distance. The sliding wear behaviour were evaluated for Al2219 alloy and hybrid composites (Al2219-3Gr with 5, 10 and 15% SiCp) at varying load (10–60 N), sliding speed (1.53–6.1 m/s) and at a constant sliding distance (5,000 m). The wear phenomenon observed were abrasion at lower applied load and sliding speed, which further exhibited delamination at higher loads [53].

3.3.2 Heat treatment

The as-cast alloys or composites are unable to possess the desired properties. The intended properties like mechanical, tribological, corrosion and so on can be carefully attained by a suitable heat treatment process. HT modifies the arrangement of microstructure resulting in phase transformation and thereby improving certain vital properties. The purpose of HT is to increase the service life of a product by improving the strength or hardness or manufacturability. HT process involves three steps: initially the specimens are heated to a suitable temperature, further it is held at a particular temperature for sufficiently longer duration so that the constituent enters into remaining solid solution. Finally, the specimens are cooled by various process like annealing, normalizing, cooling in furnace. This step-by-step process is called solution heat treatment. Ageing treatments followed by solution heat treatment are intended to control the volume fraction, size, distribution and morphology of the microconstituents [55].

3.3.2.1 Wear behaviour of alloy

A brief comparison of the as-cast and HT alloy is studied in this section. The effects of HT on wear behaviour of AA7009 alloy (as-cast and HT) were evaluated by Rao et al. [5]. The HT specimen exhibited improved wear resistance than as-cast alloy. Among the HT specimen, the specimen aged at 6 h showed superior wear behaviour. Further ageing (8 h) reduced the hardness leading to excessive wear rate. Savaskan [39] conducted a series of tests (microstructure, dimensional stability, tensile and wear test) on as-cast and HT ZA-25 alloy. He concluded that a stabilizing HT is necessary for stability in dimension. Wear volume loss increased after HT, but this was due to the presence of excess Cu content (3%). Murphy and Savaskan [40] justified the importance of HT on ZA-27 alloys by conducting wear test of as-cast and HT specimen. A steady wear behaviour was observed due to the dissolution of microconstituents (Zn-Al-Cu) and the residual stresses were relieved after HT.

Prasad [41, 42, 45, 46] performed the sliding wear test of ZA-27 alloy for as-cast and HT specimens. The specimens were solution HT at 360 °C for 12 h and aged at 180 °C for 1, 8 and 28 h. After heat treatment and ageing, the specimens were quenched in water at room temperature [41, 42]. An improvement in wear behaviour was observed with the specimen aged at 8 h over other specimens as the uniform distribution of microconstituents was observed. Further, ageing to 16 h led to coarsening of microconstituents resulting in reduction of wear resistance due to over ageing. Babic et al. [43] evaluated the sliding wear behaviour of as-cast and HT ZA-27 alloy. The specimens were HT at 370 °C for 3 and 5 h followed by water quenching at room temperature. Later, the specimens were naturally aged for 34 days. The specimen HT at 5 h exhibited superior friction co-efficient and wear behaviour over the remaining two specimens, as complete dissolution of dendritic structure was observed. The uniform distribution of microconstituents and reduced cracking tendency contributed to the higher wear resistance of the HT specimen [41–43, 45, 46]. Jovanovic et al. [56] compared the dry sliding wear resistance of ZA-27 alloy in as-cast, HT condition (370 °C for 3 h) followed by water quenching and HT followed by slow cooling. Specimens that were HT and quenched exhibited superior wear resistance and lower operating temperature than other two specimens.

3.3.2.2 Wear behaviour of SiCp reinforced MMCs

Several researchers [57–61] evaluated the sliding wear behaviour of matrix and SiCp reinforced composite specimen in both as-cast and HT condition. The effect of HT and ageing on composite was the improvement in wear resistance due to dissolution of microconstituents and relieving of residual stresses [5, 57–61], co-efficient of friction and seizure pressure [5].

3.3.2.3 Wear behaviour of Gr reinforced MMCs

Akhlaghi and Zare-Bidaki [28] evaluated the dry sliding wear behaviour of HT Al2024/Gr with graphite ranging from 5 to 20 wt.%. The comparison was made with the base alloy and concluded that composites were superior in friction co-efficient and wear resistance. The addition of graphite (more than 5 wt.%) was not recommended as the mechanical properties deteriorated resulting in delamination of composite specimen. Das and Prasad [62] compared the sliding wear behaviour of HT LM13/Gr and LM30/Gr composites with the base alloy and as-cast composites. HT composites exhibited excellent wear and seizure resistance than other specimen. Lin et al. [63] compared the sliding wear behaviour of T6 HT Al6061/Gr, annealed Al6061/Gr and base alloy. The conclusion revealed that T6 HT specimen was superior over the other specimen in dry sliding conditions.

3.3.2.4 Wear behaviour of Hybrid MMCs

Kumar and Dhiman [34] heat treated the hybrid MMCs (Al7075/SiCp-Gr) to evaluate the sliding wear behaviour. The improvement in wear resistance was observed due to work hardening, formation of iron oxide and crushing of reinforcements on surface of composite specimen. Above transition speed and load, wear resistance decreased due to reduction in thickness of MML. The transition speed and load were higher for composite than alloy specimen. Reinforcing SiCp and Gr in Al7075 exhibited higher wear resistance with no seizure in composite while seizure was observed in alloy specimen. Ravindran et al. [35, 37] evaluated the sliding wear behaviour of HT Al2024/SiCp-Gr composites. Inclusion of 5 wt.% of Gr exhibited superior wear resistance along with reduced friction coefficient. The formation of a uniform graphite layer helped to reduce the wear and friction coefficient. Guo and Tsao [38] evaluated the dry sliding wear behaviour of T6 HT Al/SiCp-Gr-Ni hybrid composites. The uncoated graphite particles showed agglomeration while graphite particles coated with nickel showed uniform distribution. Hardness, coefficient of thermal expansion and fracture energy were reduced as the wt.% of graphite increased [38]. The smeared Gr particles bonded on the surface of the specimen and there was no increase in thickness of the layer as Gr wt.% increased. The wear debris was smaller in size with an increase in the Gr content. Baradeswaran and Perumal [64] evaluated the dry sliding wear behaviour of T6 HT Al7075/Al_2O_3-Gr composite, with graphite ranging from 5 to 20 wt.%. HT composites with 5 wt.% of graphite exhibited excellent wear and seizure resistance than the base alloy and other compositions. Velmurugan et al. [65] evaluated the sliding wear behaviour of Al6061/SiCp-Gr composites. The composites were T6 HT and further aged for different ageing duration (4, 6 and 8 h). The reduction in internal stress and improvement in wear behaviour was directly related to ageing duration. Kiran et al. [66] evaluated the sliding wear behaviour of HT ZA-27/SiCp/Gr composites using Taguchi technique. The results revealed an improvement in wear resistance after HT and inclusion of 9 wt.% of SiCp. The reason for improvement was the formation of ceramic mixed mechanical layer (CMML) [66] on the subsurface of the specimen. Table 3.1 shows the overview of various matrix, reinforcement used and fabrication techniques followed by various researchers and also various tests conducted.

Table 3.1: Overview of various matrix, reinforcements, fabrication methods and tests.

Ref.	Matrix	Reinforcement			Fabrication methods	Experiments	
		Material	Size (μm)	Wt . or vol%		Mechanical	Wear
1	ZA-27	SiCp	20–30	1, 3, 5 wt.%	Liquid metallurgy	–	Pin-on-disc
2	Al2014	SiCp	50, 100	10 wt.%	Stir casting	–	Pin-on-disc, Taguchi
4	Al2024	SiCp		20 vol%	Powder metallurgy	Hardness	–
6	ZA-27	SiCp	100–150	1, 3, 5 wt.%	Stir casting	Tensile hardness Impact	–
7	AA7075	SiCp	40–150	5–25 vol%	Powder metallurgy	–	Pin on roller, Central composite design
12	ZA-37.5	SiCp	60–100	10 wt.%	Stir casting	Density Hardness	Pin-on-disc
13	Al7075	Al_2O_3 Gr (Hybrid)	16 –	2, 4, 6, 8 5 wt.%	Stir casting	Hardness Tensile Compression 3 point bending	Pin-on-disc
15	ZA-37.5	SiCp	60–100	10 wt.%	Stir casting	Density Hardness	Abrasive wear (20 μm emery paper) Pin-on-disc
16	Pure Al	SiCp	14	20 vol%	Vacuum pressure infiltration	–	Pin-on-disc
18	ZA-27	SiCp	5, 20, 80	5, 10, 15 vol%	–	Microhardness	Block-on-ring
22	A 356	SiCp	13.9	20 vol%	–	Hardness Compression Density	Block-on-ring

Table 3.1: (continued)

Ref.	Matrix	Reinforcement			Fabrication methods	Experiments	
		Material	Size (μm)	Wt . or vol%		Mechanical	Wear
24	Al7075	SiCp	2.3	10, 40 wt.%	–	–	Pin-on-disc
28	Al2024	Gr	106–150	5, 10, 15, 20 wt.%	Powder metallurgy	–	Pin-on-disc
30	ZA-27	Gr	30	2 wt.%	Compocasting	Density Hardness	Block on disc
32	LM 25	SiCp Gr (Hybrid)	10–20 70–80	1.25, 2.5, 3.75, 5 wt.%	Stir casting	Hardness	Pin-on-disc
35	Al2024	SiCp Gr (Hybrid)	43–53 43–60	5 0, 5, 10 wt.%	Powder metallurgy	–	Pin-on-disc Full factorial design
38	Al6061	SiCp Gr (Hybrid)	45 8	10 2, 5, 8 vol%	Semi solid powder densification	Hardness Coefficient of thermal expansion	Vane on disc
51, 52	Al2219	SiCp Gr (Hybrid)	25 45	15 3	Liquid metallurgy	Microhardness (subsurface)	Pin-on-disc Taguchi (52)
66	ZA-27	SiCp Gr (Hybrid)	25 45	3, 6, 9 3	Stir casting	Hardness Tensile	Pin-on-disc Taguchi
67	A6061	Alumina	30 36 24	10 15 20	–	Density Hardness	Moving-pin method Conventional Pin-on-disc method
68	Al-Si alloy	Si particles	35 to 50	18.5	–	Hardness	Pin-on-disc (Boundary lubricated)
69	A356	SiCp MoS_2	30 5		Friction stir processing	Microhardness	Pin-on-disc
70	A356	Al2O3f SiCp		20	Vacuum extraction method	Hardness	Pin-on-disc

Yang [67] evaluated the dry sliding wear behaviour of A6061/alumina composite by moving-pin and conventional pin-on-disc method. The results were compared with Archard's equation and the deviations in theoretical and experimental values were evaluated. Chen and Alpas [68] evaluated the boundary lubricated condition of Al alloy with 18.5% Si particles. A light load of 0.5 N and sliding speed of 50 m/s was applied resulting in no measurable mass loss till 6 × 10^5 cycles and no evidence of debris exhibiting ultra mild wear. The exposed Si particles took the entire applied load preventing the alloy from further removal of material. The subsurface deformation were evaluated by Alidokht et al. [69] and Wang et al. [70] which revealed higher wear resistance due to formation of layer and increased hardness along the depth that is normal to the wear surface. A comprehensive evaluation of various parameters that affected the wear behaviour were reviewed by Deuis et al. [71].

The wear surface of as-cast and HT (8 h at 370 °C) ZA-27 specimens is shown in Figure 3.1 with operating conditions of applied load 15 N, sliding speed 0.63 m/s and sliding distance of 1,000 m. The wear surface of as-cast specimen shows deep grooves while that of HT specimen exhibits fine grooves.

3.4 Wear behaviour of MMCs by design of experiments

The general practice is to evaluate the wear behaviour by varying the load, speed and sliding distance. The bulk results obtained by the experimental work help us to know the wear behaviour of the test specimen. The drawback of the experimental work is, it lacks to provide clear information on the results, that is the significance of each parameter on the wear behaviour is not clearly revealed. The statistical analysis based on Design of Experiments (DOE) highlights the significant parameters and also the interaction of parameters that affects the wear behaviour with limited experimental work. Wear behaviour is a complex phenomenon where a number of factors determine the output performance. The majority of experimental work on hybrid composites incorporating Gr and SiCp as reinforcements has shown an improvement in wear resistance. But the contribution of various parameters on the wear behaviour and performance of the specimen like reinforcement percentage, volume, size and shape, applied load, sliding speed, sliding distance and so on are identified by the help of statistical analysis.

3.4.1 Wear behaviour of MMCs

Kumar and Balasubramanian [17] evaluated the dry sliding wear behaviour of AA7075/SiCp composites using five-level central composite rotatable design matrix. Higher wear resistance were experienced by the specimen with a combination of large particle size and higher volume fraction of reinforcement. At higher load, finer particles were pulled out, while coarser particles experienced wear and fracture.

Sahin [23] evaluated the abrasive wear behaviour of Al2024 reinforced with SiCp by Taguchi technique. The results revealed that the significant parameter that affected

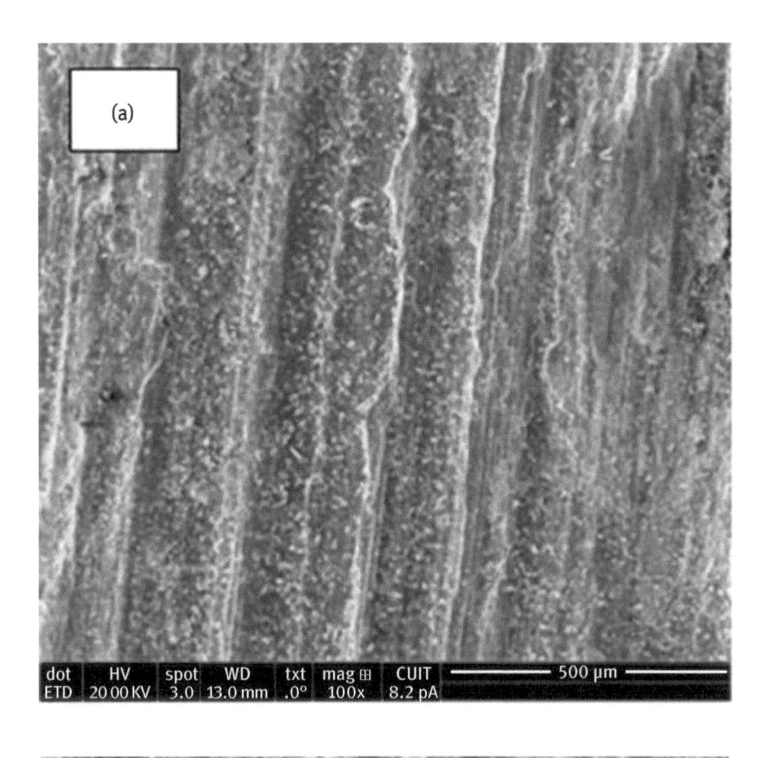

Figure 3.1: ZA-27 alloy (a) As-cast (b) HT at 15 N, 0.63 m/s, 1,000 m.

wear behaviour was abrasive grain size followed by reinforcement size. Applied load and sliding distance were the other parameters that least affected the wear behaviour. Sahoo and Pradhan [72] evaluated the effect of process parameters like cutting speed, feed, depth of cut and surface roughness in Al/SiCp MMCs under dry sliding condition by Taguchi technique. The prime wear mechanisms were abrasion followed by adhesion.

3.4.2 Wear behaviour of HMMCs

The DOE-based Central Composite Design (CCD) technique was used by Suresha and Sridhara [26, 31, 32, 36] to evaluate the sliding wear behaviour and friction coefficient of composites and hybrid composites. Sliding distance was the dominant factor which signifies that increase of sliding distance led to increase in wear loss. As sliding speed increases, wear resistance increases due to reduction in friction coefficient and as load was increased, wear resistance deteriorated. Friction coefficient was dependent on load and sliding speed, while sliding distance and percentage reinforcement had least effect. Full factorial design was used by Kumar and Dhiman [34] and Ravindran et al. [35] to find the significant factor affecting the sliding wear behaviour of Al7075 [34] and Al2024 [35] hybrid composites respectively. The significance of applied load was more compared to sliding speed and sliding distance [34, 66]. For aluminium hybrid composite, the influence of sliding distance was more compared to sliding speed and applied load [35]. Ravindran et al. [37] used Taguchi method to evaluate the dry sliding wear behaviour of Al2024 hybrid composites. The most important factor was sliding distance followed by speed and applied load. Basavarajappa et al. [52] studied the sliding wear behaviour of Al2219-based hybrid MMCs using Taguchi technique. The wear behaviour was explained on the basis of wear mechanism. The significance of sliding distance was more for both SiCp and hybrid composite followed by applied load and sliding speed.

3.5 Conclusions

The conclusions that can be drawn from the review of work by several researchers are that reinforcements and heat treatment play an important role in the performance of the components. The various fabrication methods tested revealed that stir casting method was the least expensive compared to other methods with uniform distribution of reinforcements. Mechanical properties can be improved by inclusion of SiCp in varying percentages. In order to evade the deterioration of mechanical properties, addition of Gr particles should be limited below 5 wt.%. Wear behaviour improves after inclusion of reinforcements. Hard particles (SiCp, Al_2O_3, TiC, B_4C) restrict the further exposure of base alloy while solid lubricant (Gr, MoS_2) forms a layer thereby restricting the base alloy to wear. Heat treatment results in the uniform distribution of microconstituents present in the base alloy and relieve the internal stresses produced during casting. Inclusion of more than one type of reinforcement is advantageous for tribological applications. The benefits of statistical analysis are that the significance of varying parameters that affects the performance of the specimen can be drawn with limited number of experiments.

References

[1] Sharma SC, Girish BM, Kamath R, Satish BM. Effect of SiC particle reinforcement on the unlubricated sliding wear behaviour of ZA-27 alloy composites. Wear 1997;213:33–40.

[2] Sahin Y. Optimization of testing parameters on the wear behaviour of metal matrix composites based on Taguchi method. Mater Sci Eng A 2005;24;1–8.

[3] Balasivanandha Prabhu S, Karunamoorthy L, Kathiresan S, Mohan B. Influence of stirring speed and stirring time on distribution of particles in cast metal matrix composite. J Mater Process Technol 2006;171(2):268–73.

[4] Abarghouie SM, Reihani SM. Aging behaviour of a 2024 Al alloy-SiCp composite. 2010;31(5):2368–74.

[5] Rao RN, Das S, Mondal DP, Dixit G. Effect of heat treatment on the sliding wear behaviour of aluminium alloy (Al–Zn–Mg) hard particle composite. Tribol Int 2010;43(1–2):330–9.

[6] Seah KH, Sharma SC, Rao PR, Girish BM. Mechanical properties of as-cast and heat-treated ZA-27/silicon carbide particulate composites. Mater Des 1995;16(5):277–81.

[7] Kumar S, Balasubramanian V. Developing a mathematical model to evaluate wear rate of AA7075 / SiC p powder metallurgy composites. Wear 2008;264:1026–34.

[8] Sadık B, Atik E. Tribological properties of journal bearings manufactured from particle reinforced Al composites. Mater Des 2009;30(4):1381–5.

[9] Prasad BK. Effectiveness of an externally added solid lubricant on the sliding wear response of a zinc–aluminium alloy, its composite and cast iron. Tribol Lett 2005;18(2):135–43.

[10] Sozhamannan GG, Balasivanandha Prabu S. An experimental investigation of the interface characteristics of aluminium/silicon carbide. J Alloys Compd 2010;503(1):92–5.

[11] Wilson S, Alpas AT. Effect of temperature on the sliding wear performance of A1 alloys and Al matrix composites. Wear 1996;196:270–8.

[12] Prasad BK. Sliding wear response of a zinc-based alloy and its composite and comparison with a gray cast iron: Influence of external lubrication and microstructural features. Mater Sci Eng: A 2005;392(1–2):427–39.

[13] Baradeswaran A, Elaya Perumal A. Study on mechanical and wear properties of Al 7075/Al2O3/graphite hybrid composites. Composites: Part B 2014;464–71.

[14] Vieira AC, Sequeira PD, Gomes JR, Rocha LA. Dry sliding wear of Al alloy/SiCp functionally graded composites: Influence of processing conditions. Wear 2009;267:585–92.

[15] Prasad BK, Modi OP, Khaira HK. High-stress abrasive wear behaviour of a zinc-based alloy and its composite compared with a cast iron under varying track radius and load conditions. Mater Sci Eng: A 2004;381(1–2):343–54.

[16] Lu D, Gu M, Shi Z. Materials transfer and formation of mechanically mixed layer in dry sliding wear of metal matrix composites against steel. Tribol Lett 1999;6:57–61.

[17] Kumar S, Balasubramanian V. Effect of reinforcement size and volume fraction on the abrasive wear behaviour of AA7075 Al/SiCp P/M composites – A statistical analysis. Tribol Int 2010;43(1–2):414–22.

[18] Tjong SC, Chen F. Wear Behaviour of As-Cast ZnAl27/SiC particulate metal-matrix composites under lubricated sliding condition. Wear 1997;28:1951–5.
[19] Prasanna kumar M, Sadashivappa K, Prabhukumar GP, Basavarajappa S. Dry sliding wear behaviour of garnet particles reinforced zinc-aluminium alloy metal matrix composites. Mater Sci 2006;12(3):209.
[20] Sahoo AK, Pradhan S. Modeling and optimization of Al/SiCp MMC machining using Taguchi approach. Meas 2013;46(9):3064–72.
[21] Kwok JK, Lim SC. High-speed tribological properties of some Al/SiCp composites: I. Frictional and wear-rate characteristics. Compos Sci Technol 1999;59:55–63.
[22] Wilson S, Alpas AT. Wear mechanism maps for metal matrix composites. Wear 1997;212(1): 41–9.
[23] Sahin Y. Optimization of testing parameters on the wear behaviour of metal matrix composites based on the Taguchi method. Mater Sci Eng A 2005;408:1–8.
[24] Venkataraman B, Sundararajan G. Correlation between the characteristics of the mechanically mixed layer and wear behaviour of aluminium, Al-7075 and Al-MMCs. Wear 2000;245:22–38.
[25] Seah KH, Sharma SC, Girish BM. Effect of artificial ageing on the hardness of cast ZA-27/graphite particulate composites. Mater Des 1995;16(6):337–41.
[26] Suresha S, Sridhara BK. Effect of silicon carbide particulates on wear resistance of graphitic aluminium matrix composites. Mater Des 2010;31(9):4470–7.
[27] Girish BM, Prakash KR., Satish BM, Jain PK, Prabhakar P. An investigation into the effects of graphite particles on the damping behaviour of ZA-27 alloy composite material. Mater Des 2011;32(2):1050–6.
[28] Akhlaghi F, Zare-Bidaki A. Influence of graphite content on the dry sliding and oil impregnated sliding wear behaviour of Al 2024 – graphite composites produced by in situ powder metallurgy method. Wear 2009;266:37–45.
[29] Seah KH, Sharma SC, Girish BM, Lima SC. Wear characteristics of as-cast ZA-27/graphite particulate composites. Mater Des 1996;17(2):63–7.
[30] Babic M, Slobodan M, Dragan D, Jeremic B, Ilija B. Tribological behaviour of composites based on ZA-27 alloy reinforced with graphite particles. Tribol Lett 2010;37:401–10.
[31] Suresha S, Sridhara BK. Friction characteristics of aluminium silicon carbide graphite hybrid composites. Mater Des 2012;34:576–83.
[32] Suresha S, Sridhara BK. Wear characteristics of hybrid aluminium matrix composites reinforced with graphite and silicon carbide particulates. Compos Sci Technol 2010;70(11):1652–9.
[33] Riahi A, Alpas A. The role of tribo-layers on the sliding wear behaviour of graphitic aluminum matrix composites. Wear 2001;251(1–12):1396–407.
[34] Kumar R, Dhiman S. A study of sliding wear behaviors of Al-7075 alloy and Al-7075 hybrid composite by response surface methodology analysis. Mater Des 2013;50:351–9.
[35] Ravindran P, Manisekar K, Narayanasamy P, Selvakumar N, Narayanasamy R. Application of factorial techniques to study the wear of Al hybrid composites with graphite addition. Mater Des 2012;39:42–54.
[36] Suresha S, Sridhara BK. Effect of addition of graphite particulates on the wear behaviour in aluminium–silicon carbide–graphite composites. Mater Des 2010;31(4):1804–12.
[37] Ravindran P, Manisekar K, Narayanasamy R, Narayanasamy P. Tribological behaviour of powder metallurgy-processed aluminium hybrid composites with the addition of graphite solid lubricant. Ceram Int 2013;39(2):1169–82.
[38] Guo ML, Tsao CA. Tribological behaviour of self-lubricating aluminium/SiC/graphite hybrid composites synthesized by the semi-solid powder-densification method. Compos Sci Technol 2000;60:65–74.

[39] Savaskan T. Mechanical properties and lubricated wear of ZN-25Al-based alloys. Wear 1987; 116:211–24.
[40] Murphy S, Savaskan T. Comparitive wear behaviour of Zn-Al-based alloys in an automotive engine application. Wear 1984;98:151–61.
[41] Prasad BK. Influence of heat treatment parameters on the lubricated sliding wear behaviour of a zinc-based alloy. Wear 2004;257(11):1137–44.
[42] Prasad BK. Microstructure and tensile property characterization of a nickel-containing zinc-based alloy: effects of heat treatment and test conditions. Mater Sci Eng A 2000;277:95–101.
[43] Babic M, Mitrovic S, Jeremic B. The influence of heat treatment on the sliding wear behaviour of a ZA-27 alloy. Tribol Int 2010;43(1–2):16–21.
[44] Babic M, Vencl A, Mitrovic S, Boboc I. Influence of T4 heat treatment on tribological behaviour of ZA27 alloy under lubricated sliding condition. Tribol Lett 2009;125–34.
[45] Prasad BK. The effect of heat treatment on sliding wear behaviour of a zinc-based alloy containing nickel and silicon. Tribol Lett 2003;15(3):333–41.
[46] Prasad BK. Effect of microstructure on the sliding wear performance of a Zn–Al–Ni alloy. Wear 2000;240:100–12.
[47] Sharma SC, Sastry S, Krishna M. Effect of aging parameters on the micro structure and properties of ZA-27 / aluminite metal matrix composites. J Alloys Compd. 2002;346: 292–301.
[48] Ranganath G, Sharma SC, Krishna M, Muruli MS. A study of mechanical properties and fractography of ZA-27 / titanium-dioxide metal matrix composites. J Mater Eng Perform 2002;11:408–13.
[49] Sharma SC, Girish BM, Somashekar DR, Kamath R, Satish BM. Mechanical properties and fractography of zircon-particle-reinforced ZA-27 alloy composite materials. Compos Sci Technol 1999;59:1805–12.
[50] Prasad BK, Yegneswaran H, Patwardan AK. Influence of the nature of microconstituents on the tensile properties of a zinc-based alloy and a leaded-tin bronze at different strain rates and temperatures. J. of Material Science. 1997;32:1169–75.
[51] Basavarajappa S, Chandramohan G, Mahadevan A, Thangavelu M, Subramanian R, Gopalkrishnan P. Influence of sliding speed on the dry sliding wear behaviour and the subsurface deformation on hybrid metal matrix composite. Wear 2007;262:1007–12.
[52] Basavarajappa S, Chandramohan G, Davim JP. Application of Taguchi techniques to study dry sliding wear behaviour of metal matrix composites. Mater Des 2007;28:1393–8.
[53] Basavarajappa S, Chandramohan G, Mukund K, Ashwin M, Prabhu M. Dry sliding wear behaviour of Al 2219/SiCp-Gr hybrid metal matrix composites. JMEPEG 2006;15:668–74.
[54] Basavarajappa S, Chandramohan G. Dry sliding wear behaviour of metal matrix composites: A Statistical approach. JMEPEG 2006;15:656–60.
[55] Heat treating, Volume 4, ASM Handbook
[56] Jovanivic MT, Bobic I, Djuric B, Grahovac N, Ilic N. Microstructural and sliding wear behaviour of a heat-treated zinc-based alloy. Tribol Lett 2007;25(3):173–84.
[57] Maleque MA, Karim MR. Wear behaviour of as-cast and heat treated triple particle size SiC reinforced aluminum metal matrix composites. Ind Lubr Tribol 2009;61(2):78–83.
[58] Li XY, Tandon KN. Subsurface microstructures generated by dry sliding wear on as-cast and heat treated Al metal matrix composites. Wear 1997;203–204:703–8.
[59] Kiourtsidis GE, Skolianos S. Wear behaviour of artificially aged AA2024/40 μm SiCp composites in comparison with conventionally wear resistant ferrous materials. Wear 2002;253(9–10):946–56.
[60] Dasgupta R, Meenai H. SiC particulate dispersed composites of an Al–Zn–Mg–Cu alloy: Property comparison with parent alloy. Mater Charact 2005;54(4–5):438–45.

[61] Yamanoğlu R, Karakulak E, Zeren A, Zeren M. Effect of heat treatment on the tribological properties of Al–Cu–Mg/nano SiC composites. Mater Des 2013;49:820–5.
[62] Das S, Prasad SV. Microstructure and wear of cast (Al-Si alloy)-graphite composites. Wear 1989;133(1):173–87.
[63] Lin CB, Chang RJ, Weng WP. A study on process and tribological behaviour of Al alloy/Gr.(p) composite. Wear 1998;217(2):167–74.
[64] Baradeswaran A, Elaya Perumal A. Wear and mechanical characteristics of Al 7075/graphite composites. Composites Part: B. 2014;56:472–6.
[65] Velmurugan C, Subramanian R, Thirugnanam S, Anandavel B. Investigation of friction and wear behaviour of hybrid aluminium composites. Ind Lubr Tribol 2012;64(3):152–63.
[66] Kiran TS, Prasanna Kumar M, Basavarajappa S, Viswanatha BM. Dry sliding wear behaviour of heat treated hybrid metal matrix composite using Taguchi techniques. Mater Des 2014;63:294–304.
[67]. Yang LJ. A methodology for the prediction of standard steady-state wear coefficient in an aluminium-based matrix composite reinforced with alumina particles. J Mater Process Technol 2005;162–163:139–48.
[68] Chen M, Alpas AT. Ultra-mild wear of a hypereutectic Al–18.5 wt.% Si alloy. Wear 2008;265:186–95.
[69] Alidokht SA, Abdollah-Zadehn A, Assadi H. Effect of applied load on the dry sliding wear behaviour and the subsurface deformation on hybrid metal matrix composite. Wear 2013;305:291–8.
[70] Wang YQ, Afsar AM, Jang JH, Han KS, Song JI. Room temperature dry and lubricant wear behaviors of Al2O3f/SiCp/Al hybrid metal matrix composites. Wear 2010;268:863–70.
[71] Deuis RL, Subramanian C, Yellup JM. Dry slip wear of aluminium composites – a review. Compos Sci Technol 1997;57(4):415–35.
[72] Sahoo AK, Pradhan S. Modeling and optimization of Al/SiCp MMC machining using Taguchi approach. Meas 2013;(46):3064–72.

G. Gautam, N. Kumar, A. Mohan, and S. Mohan

4 Tribology of aluminium matrix composites

4.1 Introduction

Tribology is characterized by the study of wear, friction and lubrication. Components in automobile, aerospace and marine sector undergo relative motions and suffer high amount losses in the forms of materials and energy due to wear and friction. These losses could be minimized either by the replacement of components by new materials or by the application of suitable surface treatment on components [1]. In the last decade, aluminium matrix composites (AMCs) have been paid lot of attention by scientific community. Composites broadly consist of a continuous phase termed as matrix and discontinuous one as reinforcement. On the basis of matrix, composites are defined as metal matrix composites (MMCs), ceramic matrix composites (CMCs) and organic matrix composites (OMCs). AMCs are the member of MMCs and these are one of the important engineering materials used in automotive, aerospace and marine applications due to their unique combination of properties such as low density, high strength and high wear resistance at ambient and high temperature. In the AMCs, aluminium or aluminium alloy matrix is reinforced with intermetallic or ceramic particles/fibres [2–4]. The fabrication process of AMCs could be *exsitu* (reinforcement is added externally) or *insitu* (reinforcement generates within the melt by chemical reaction). However, reinforcement generated within the melt exhibits the advantages of homogeneous distribution of reinforcement in finer size, clean interface with the matrix, and thermodynamic equilibrium with matrix. These advantages help to improve the mechanical and tribological properties of AMCs [5–11]. This chapter addresses the different parameters affecting dry sliding wear and friction of AMCs with a focus on wear mechanism.

4.2 Wear of AMCs

Wear of the material is characterized as the loss of material under relative motion and it is quantified in terms of wear volume or wear rate. Wear volume is defined as the loss of material per unit sliding distance. However, wear rate is defined as the wear per unit volume per unit sliding distance. But in certain cases, it has been defined as material loss per unit time. Large number of factors affects wear performance of a system which may include primary or secondary operating parameters, environmental conditions, material properties and so on. Further, operative wear mechanisms change with different set of conditions; hence, a proper analysis of worn out surface is important. It needs to be done through roughness profile, morphology and phases present.

https://doi.org/10.1515/9783110352986-004

4.2.1 Operating parameters

Operating parameters which chiefly decide AMCs as wear material include sliding distance, applied load and sliding velocity. Effect of these parameters on wear properties has been explained in following subsections:

4.2.1.1 Sliding distance

Bulk wear or volume loss of AMCs changes linearly with the sliding distance after an initial running-in period. At short distances, the volume loss is less due to the oxidation of the composite surface and the wear track is smooth with a thin oxide layer and the debris is also mostly oxidative in nature. But as time elapses and distance moved is large, volume loss increases due to the change in wear mechanism from oxidative to oxidative-metallic and worn surface displays thick oxide layer with a deeper track [12, 13]. Debris of such regions consists of oxides with metallic particles. Figure 4.1 shows bulk wear with sliding distance for Al-Al_3Fe composite [12].

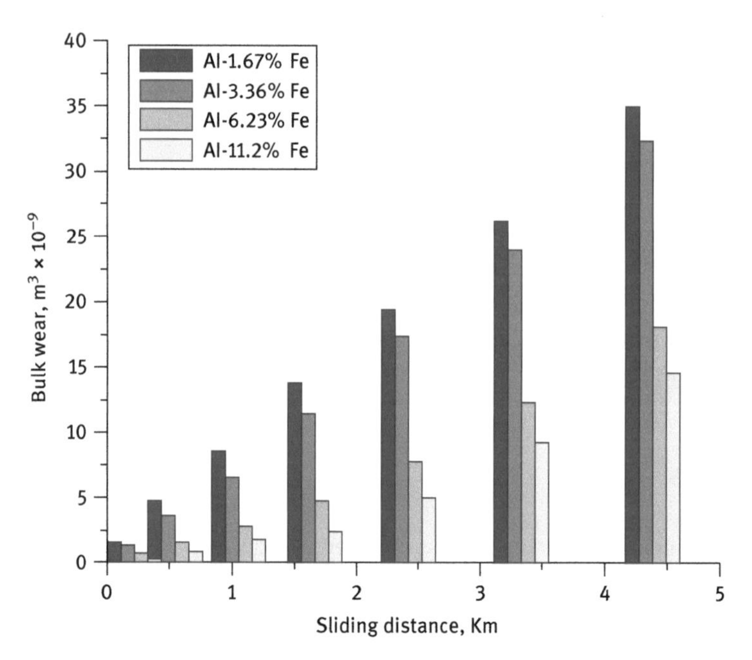

Figure 4.1: Effect of sliding distance on bulk wear for as-cast Al-Al_3Fe composites.

When abrasive conditions are consistently maintained with a fresh counter-face, a severe wear trend is observed due to the continuous protrusion of sharp hard particles in AMCs. Whereas, if this consistency is avoided and process is continued with same abrasive counter-face, mild wear behaviour is observed. In the first case, linear behaviour of wear is observed because every time AMC faces a new counter-face with sharp particles of same roughness, while in the second case, the process is continued with same counter-face and abrasive particles get blunt with distance. Thus, the grooves created are not as deep as in the first case. As a result, volume loss decreases and cumulative wear volume becomes non-linear showing less wear [14].

4.2.1.2 Applied load

In AMCs and aluminium alloys, it is common to observe increasing trend of wear rate with increase in applied load. Increase in wear rate remains less up to a certain load due to the formation of the oxide layer which reduces the contact between the surfaces. At such loads, mild/oxidative wear mechanism is observed with roughness value of 1.98 μm. As the load reaches a critical value, hike in wear rate is observed for all compositions. In this case, worn surface consists of deep grooves and oxide particles causing the wear mechanism shift from mild/oxidative to severe/oxidative-metallic or even severe/metallic at reasonably very high loads. The roughness value also shows a substantial increase from 1.98 μm to 4.00 μm [15]. Figure 4.2(a and b) shows same nature with load for wear rate and specific wear rate (wear rate per unit vol.% Al_3Zr) for AMC reinforced with Al_3Zr particles.

4.2.1.3 Sliding velocity

With sliding velocity, one observes both decreasing and increasing trends in different range of sliding velocities for aluminium matrix composites. It cannot be treated as thumb rule but generally a decreasing trend is observed at low working sliding velocities followed by a hike in wear rate after a critical value. This critical value depends on number of variables. However, responsible mechanisms are different for the cases mentioned above [16–19].

In AMCs, increase in sliding velocity promotes formation of transfer layer in the matrix region. This transfer layer acts as a protective layer on the surface during the sliding process and direct metal to metal contact is reduced, which finally decreases wear rate. The fraction of this transfer layer increases covering larger area of counteracting surface and wear rate continues to decrease with sliding velocity [16]. However, after a critical value, due to high strain rates the deformation at subsurface results crack generation in oxide film. Hence, dislodging of film takes place increasing the metal to metal contact. This overall process promotes delamination of the surface and large fragments come out of the surface and wear rate increases sharply [17]. According to other theories, temperature of worn surface increases with an increase in sliding velocity and surface softening occurs due to frictional heating. Embedded hard particles in the AMCs may come out of the surface and may act as third body at the mating interface causing increase in wear rate several folds [18]. Figure 4.3 shows a common trend of wear

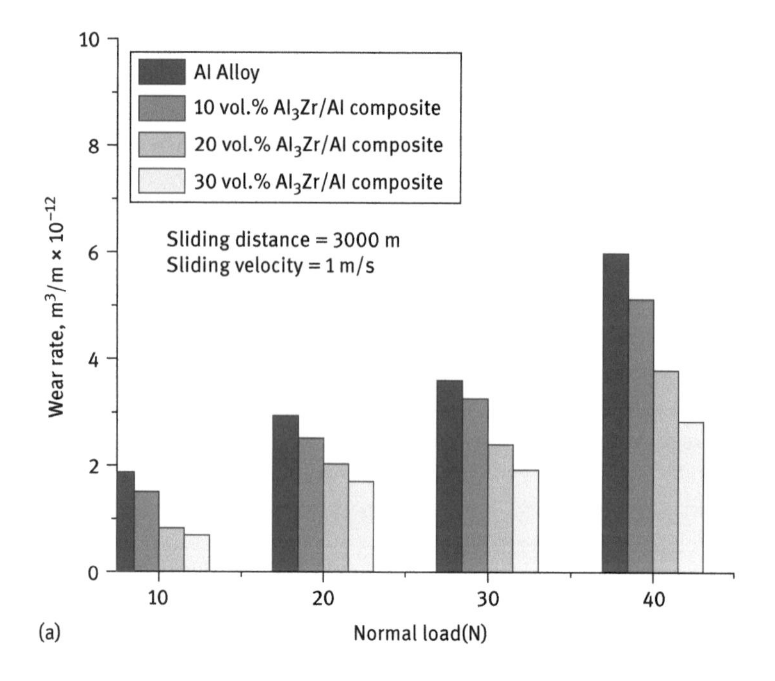

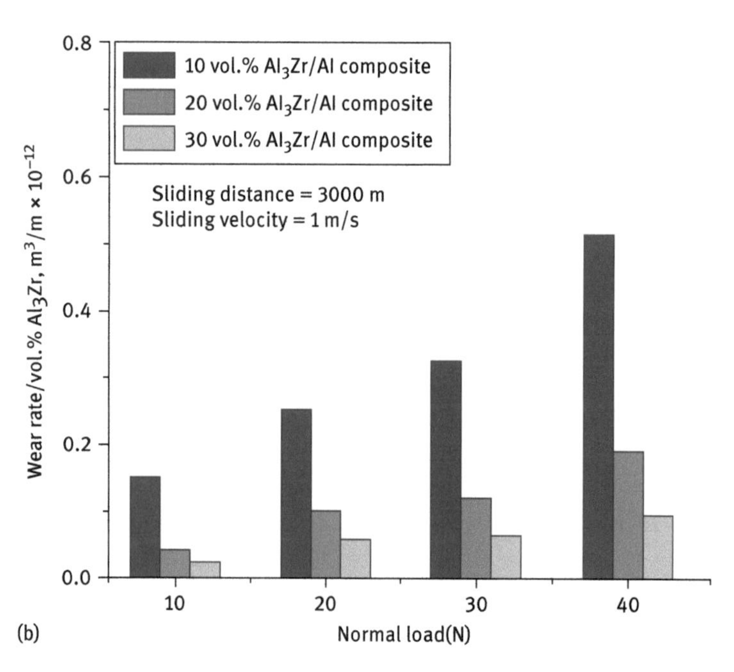

Figure 4.2: Effect of applied load on (a) wear rate, and (b) wear rate per unit vol.% Al_3Zr in alloy and composites.

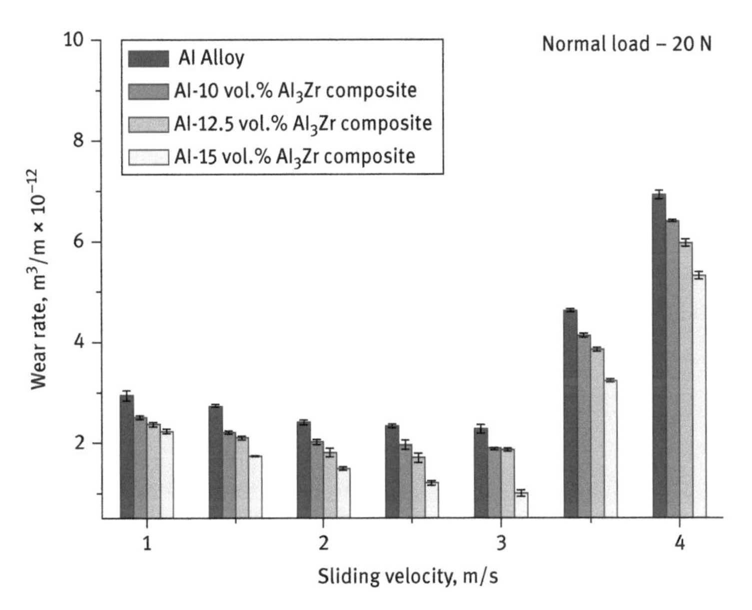

Figure 4.3: Effect of sliding velocity on wear rate.

rate with sliding velocity for *insitu* generated Al_3Zr particles reinforced aluminium matrix composite. In the initial stage, oxidation of the surface occurs which avoids the direct metal to metal contact resulting decrease in wear rate and the responsible wear mechanism is mild/oxidative. But beyond a critical value of sliding velocity, that is 3 m/s, wear rate increases rapidly due to the fragmentation of hard oxide layer which acts as a third body abrasive at the mating interface during sliding, and wear mechanism changes from mild/oxidative to severe/oxidative-metallic [19].

4.2.2 Environmental condition

4.2.2.1 Temperature

Wear rate of AMCs increases with increase in working temperature. With increase in temperature, the surface of AMCs gets softer and the hard asperities of hard counter-face penetrate soft AMC's surface which damage it to a large extent causing wear rate to increase. With the incorporation of hard particles in the AMCs, the transition temperature of mild-oxidative wear to severe-metallic wear shifts to higher values in comparison to matrix alloy [20]. This is due to the better dimensional stability of matrix provided by the high amount of hard reinforcement particles in composites [21–24].

However, typical patterns are observed in different cases. It has been observed by Gautam et al. [25] that in hybrid AMCs with different particle configurations, wear rate increases with increase in temperature followed by a dip in wear rate, and finally it again increases sharply with further increase in temperature (Figure 4.4). The hybrid AMCs consisted of large and small reinforcement particles. The matrix of AMC gets soft on increase in temperature and larger reinforcement particles protrude from the surface which breaks under applied pressure. These broken particles may be removed during sliding and cause an increase in wear rate. But, matrix of AMC gets softer with further rise in temperature and the merging of these protuberances takes place. At the same time, oxidation of the surface also increases which covers the large area of the surface and both the phenomena contribute in decreasing the wear rate. Oxide layer gets thicker and breaks but soft matrix is unable to hold the hard particles on further increase in temperature, which again increases the wear rate. Worn surface shows a severe delamination and large amount of metal flow at high temperatures. In addition to that, surface roughness value increases from 4.00 to 8.64 µm as the temperature increase, which is also in agreement with wear results.

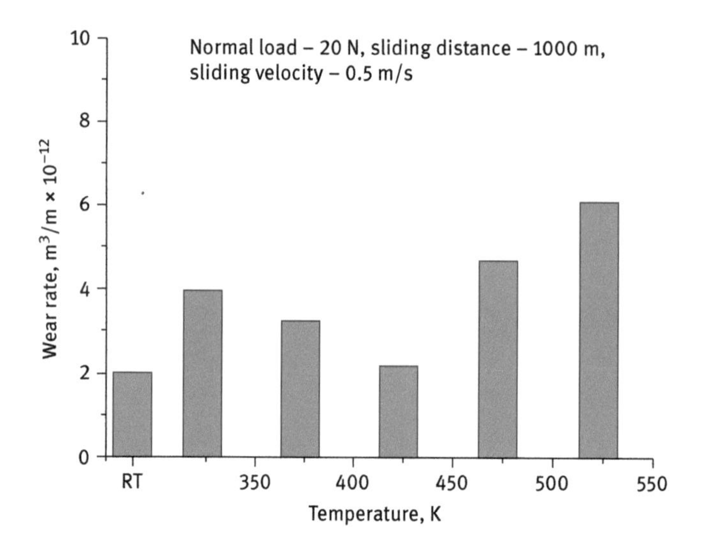

Figure 4.4: Effect of temperature on wear rate.

4.2.3 Material property

4.2.3.1 Hardness of AMCs

Wear rate is explained as the volume loss of material per unit sliding distance. It depends on the hardness of the material. According to the Archard's law of wear ($Q = kW/H$, where Q is volume loss of material per unit sliding distance, W is the applied load, H is the hardness of the wearing surface and k is the wear coefficient) [26], the wear rate is inversely proportional to the hardness of testing material.

Another study describes that the specific wear rate (SWR) is the function of the hardness of the composites and decreases with an increasing hardness of the composites [27]. SWR is explained as the wear volume of material per unit load per unit sliding distance. Higher hardness of the composites decreases the ploughing tendency during wear. Ploughing occurs due to the interaction between the hard asperity of the counter-face with a surface of composite. At low hardness, worn surface shows deep, long grooves with high value of average surface roughness (Ra). While at higher hardness, worn surface exhibits shallow and narrow grooves and surface looks smoother with less value of surface roughness (Ra).

4.2.3.2 Morphology of reinforcement

Wear of the AMCs also depends on the morphology of reinforcement phases. There are three morphologies of reinforcement phases which are generally used in AMCs, namely particle, whisker and fibre. Whiskers and fibres provide directional properties, whereas particles provide isotropic properties. Hence, among these morphologies particle reinforcement is preferred. Further, particle reinforcement has a large diameter as compared to whisker and fibre, so it prevents the initial severe wear and also decreases the steady-state wear due to the large load supporting efficiency [28].

4.2.3.3 Size of reinforcement

The size of the reinforcement also affects the wear properties of AMCs. As the size of reinforcement increases, the wear resistance of AMCs increases. The larger size of reinforcement in the AMCs acts as protecting agent and reduce the metal to metal contact. As a result, improvement in wear resistance of AMCs is achieved [29–30]. Worn surface of the AMCs with smaller size reinforcement exhibits high density of surface cracks, while AMCs consisting of large size reinforcement, smoother worn surface with less surface cracks and pits is observed [31].

4.2.3.4 Amount of reinforcement

Wear rate of AMCs decreases with an increasing amount of reinforcement phase. Figure 4.5(a–c) shows a typical case of ZrB_2 particles reinforcement. The variation in wear rate, specific wear rate (SWR) and normalized wear rate of AMCs with the volume fraction of reinforcement particles have been shown in this figure. All of them decrease with an increase in volume fraction of reinforcement particles in AMCs. The decrease in wear rate takes place due to the formation of mechanically mixed layer

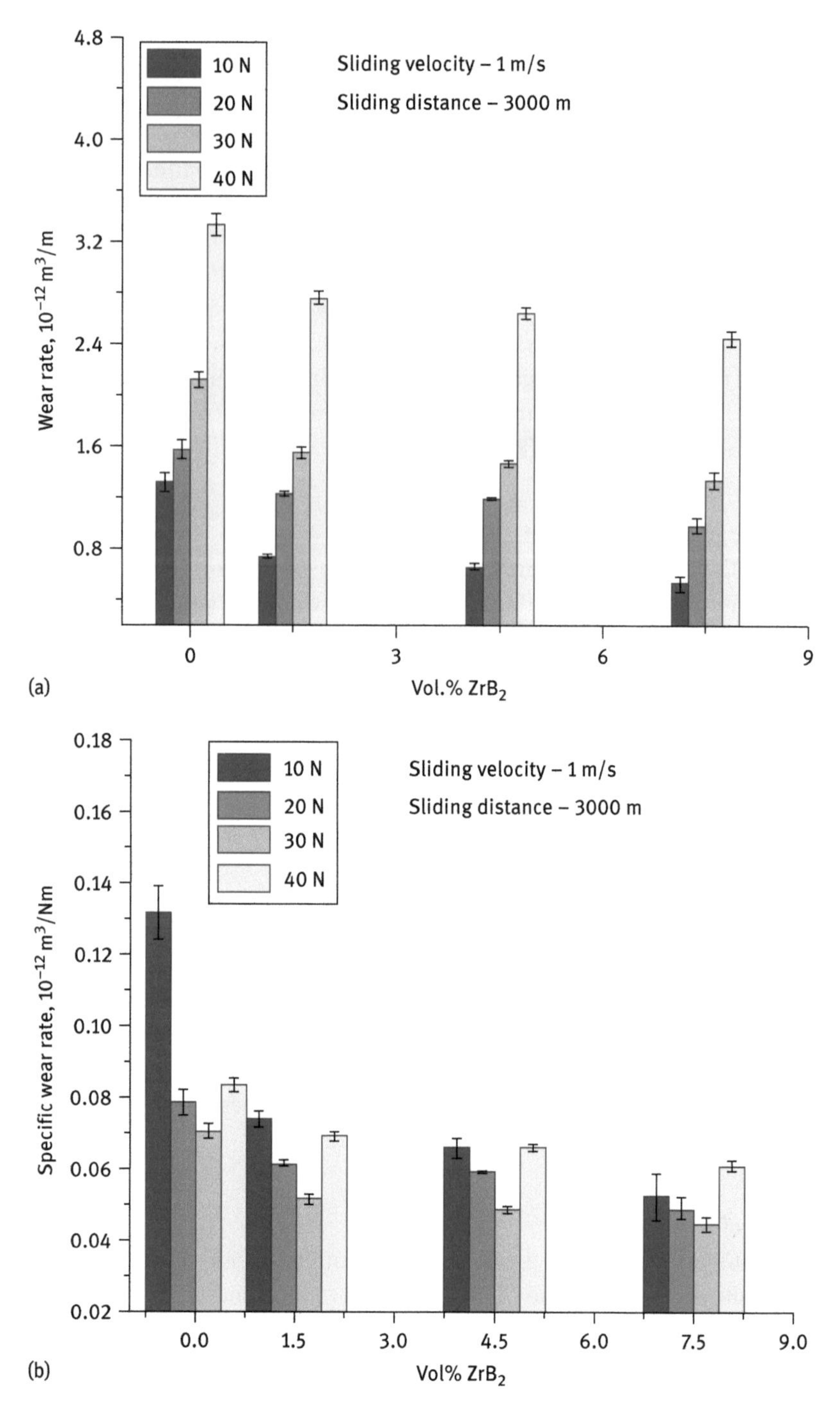

Figure 4.5: Variation of (a) wear rate, (b) specific wear rate and (c) normalized wear rate with vol% ZrB_2.

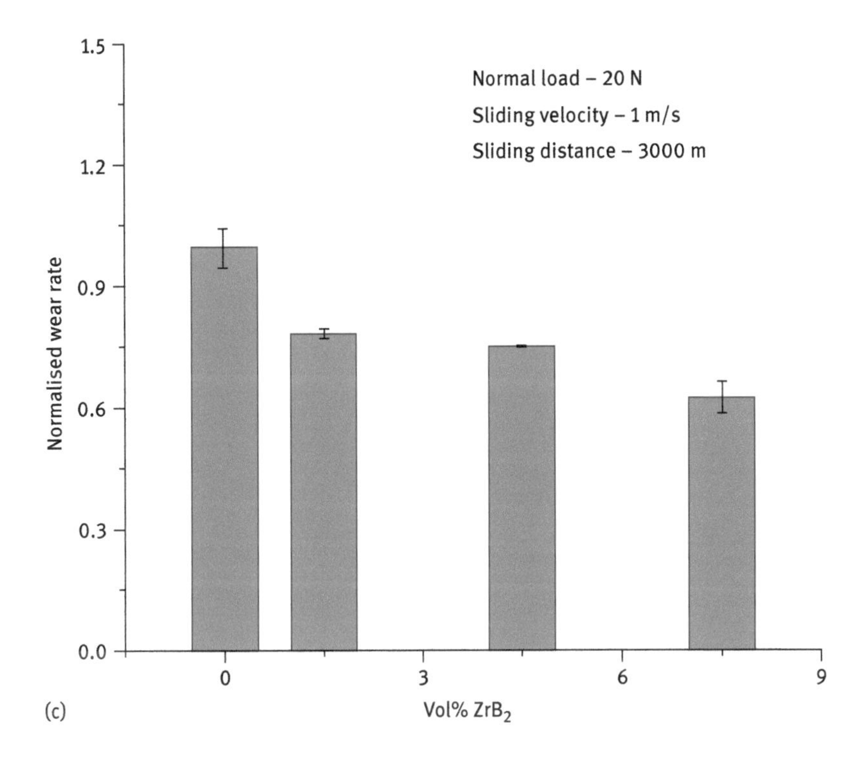

Figure 4.5: (continued)

(MML) on the surface of AMCs [18]. MML layer consists of soft aluminium matrix and hard reinforcement particle. The hardness of MML increases with an increase in the amount of reinforcement particles. MML layer restricts the loss of material from the surface and decrease the wear rate. This phenomenon is also in agreement with Archard's wear law [26]. Apart from this, the wear rate of AMCs may also decrease due to the interaction of hard reinforcement particles with counter-face. The hard reinforcement particles may interact with counter-face under sliding motion restricting the flow of material causing the wear rate to decrease. This phenomenon gets strengthened with an increase in amount of hard particles in the composites resulting in the reduced wear rate [32, 33]. Further, an increasing amount of hard reinforcement particles in AMCs also enhances grain refinement. Thus, an increment in interfacial bonding and dislocation density occurs which further improves the hardness and strength of the AMCs and contribute to decrease in wear rate [34].

4.2.3.5 Multiple reinforcements

Wear rate of AMCs significantly decreases with the incorporation of multiple reinforcements. Figure 4.6 shows the comparative study of AMCs with multiple and

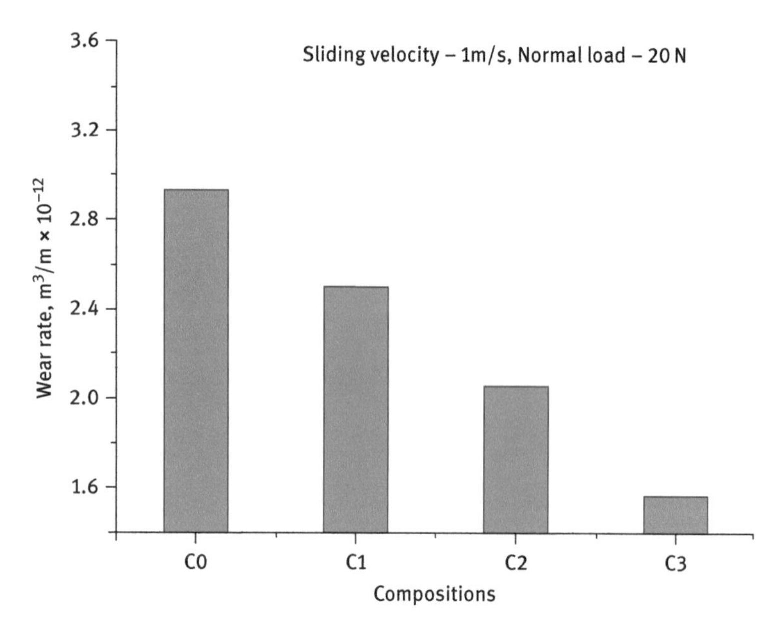

Figure 4.6: Variation in wear rate with compositions, namely C0 (AA5052 alloy), C1 (10 vol.%Al_3Zr/AA5052), C2 (3 vol.%ZrB_2/AA5052), C3 (10 vol.%Al_3Zr + 3 vol.%ZrB_2/AA5052).

single reinforcement [35]. Among all composites, AMC reinforced with multiple reinforcement shows minimum wear rate. This is due to the higher hardness of the composite provided by the multiple reinforcements than AMCs consisting of single reinforcement. This restricts the flow of material from the composite surface in a better way during sliding and wear rate decreases. Worn surfaces show the high amount of plastic deformation and broken oxide layer with higher surface roughness values in base alloy (2.30 µm) and AMCs with single reinforcement, that is Al_3Zr (1.82 µm) and ZrB_2 (1.59 µm). However, with multiple reinforcement, that is (Al_3Zr+ZrB_2), the worn surface of AMC reveals less deformation and ploughing with minimum surface roughness value of 1.40 µm.

4.2.4 Synthesis

4.2.4.1 Insitu reaction temperature and holding time

Wear properties of AMCs are also affected by the *insitu* reaction temperature during synthesis of the AMCs. With alteration in process, temperature morphology of the reinforcement changes that affects the wear properties of AMCs. Morphology of reinforcement in AMCs may have shapes such as spherical, tetragon, rod and fibre with change in temperature [36]. Wear loss also changes with changed morphology of the reinforcement; however, minimum wear loss has been observed in particle reinforced AMCs. Wear rate is also affected by holding time for *insitu* reaction and generally it decreases with an increased reaction holding time [37]. Increased *insitu* reaction holding time leads to finer reinforcement in large amount that restricts the growth of aluminium-rich grains. The final microstructure consists of finer aluminium-rich grains with finer reinforcement particles in large number. This overall process increases the hardness of the composite significantly, which restricts plastic deformation of surface thereby wear rate decreases [27, 38].

4.2.4.2 Stirring

Wear rate of AMCs is also affected by stirring applied during their synthesis. Stirring applied in the molten condition drastically reduces the size of reinforcement particles in the mushy state. Reduction in size is due to the high amount of mechanical force provided by stirring which breaks the reinforcement particles [39]. These fine reinforcement particles are incapable of supporting high loads during sliding. At times chemical reaction between atmospheric air and reinforcement particles takes place that causes poor interfacial properties. Due to weaker interfacial bonding, the dislodging of particles occurs from the surface giving rise to large craters in worn surface and wear rate increases [14].

4.2.4.3 Type of casting process

Wear of AMCs is also affected by the casing routes. Permanent mould and squeeze casting are two common processes for composite preparation through melting route. In the squeeze casting process, during solidification interatomic distance decreases due to the application of pressure which restricts the atomic movement. As a result, aluminium-rich matrix with much finer grains as compared to permanent mould process is achieved. These finer grains increase hardness of the composite and improve the wear resistance [40]. In addition, applied pressure also reduces the defects that also help in improving wear resistance of AMCs [41].

4.2.4.4 Secondary processing

Secondary processes like heat treatment, rolling and so on also affect the wear properties of AMCs. Wear rate initially decreases with ageing time in Al-Zn-Mg matrix composites due to precipitate formation. This phenomenon is controlled by ageing time and temperature both. But with further ageing, size of the precipitates grows and wear rate starts increasing. Initially with precipitation hardening of the composites takes place and helps to reduce wear rate but with precipitates coarsening or with over ageing hardness decreases and wear rate increases [42]. Mushy state rolling also improves the wear resistance of AMCs. Mushy state rolling improves the hardness by grain refinement and particle redistribution [43, 44].

4.2.5 Inclusions in mechanically mixed layer

The wear rate of AMCs decreases with the formation of mechanically mixed layer (MML). During sliding, the steel inclusion transfers from the counter-face to the surface of the AMCs and acts as an additional reinforcement which decreases the wear rate [45, 46]. Formed MML layer on the surface also acts as a protecting layer and solid lubricant [47, 48] causing wear rate of AMCs to decrease. Wear rate further decreases with increasing thickness of the MML layer [49]. However, the stability of MML layer depends on the volume fraction of reinforcement, inclusion transfer, load and so on. With less volume fraction of particles, MML remains stable only in low load range. To have stable MML at high loads, large volume fraction of particles is required [50]. Formation of stable MML helps in delaying mild to severe wear transition in AMCs [51].

4.3 Friction of AMCs

During relative motion, friction is the main cause of energy loss. It is quantified in terms of friction force and coefficient of friction (COF) [52]. Coefficient of friction (COF) is defined as the friction force per unit applied normal load. Though coefficient of friction depends on material pair but like wear, it also depends on the test operating conditions like sliding distance, applied loads, sliding velocity, material properties, amount of reinforcement and environmental conditions. Other parameters such as *insitu* reaction holding time and mechanical mixed layer, and so on also affect the COF. COF consist of four components, namely asperity deformation, adhesion, ploughing and third body [1]. The variation in the sliding distance during sliding also affects the COF of AMCs and it shows the fluctuating behaviour. This is due to the protruding hard particles coming into contact with counter face [33]. COF may exhibit decreasing trend with applied load when hard particles get merged into AMC [15].

Coefficient of friction does not have any fixed pattern. Formation of MML in AMCs affects coefficient of friction significantly. Initially at low load, decrease in COF with applied load occurs due to the formation of MML which provides a smooth surface. But at high load the MML layer may develop cracks and hard particles may come out at the interface which leads to increase in COF [18]. However, COF may show the fluctuating behaviour with applied load which is due to periodic formation and removal of oxides or at times it may be due to the occurrence of different complex reactions during the sliding [53, 54]. Rise in temperature takes place with increase in sliding velocity that promotes oxide formation on the surface of AMCs. This process increases the overall hard particle contribution in MML causing COF to rise [18, 33]. In certain situations, asperity junctions also increase with temperature rise contributing to COF [17, 55]. Large amount of reinforcement particles in the AMCs is also one of the factors for increased COF [18, 33].

Working temperature is also important in friction studies. Coefficient of friction of AMCs increases with an increase in test temperatures (Figure 4.7). With temperature rise, softening of composite surface occurs and oxide formation is also increased. These hard oxide particles act as protuberances and increase the coefficient of friction [20, 25]. In addition to that, adherence between counter disk and pin sample may also increase with test temperature and contributes to COF [56].

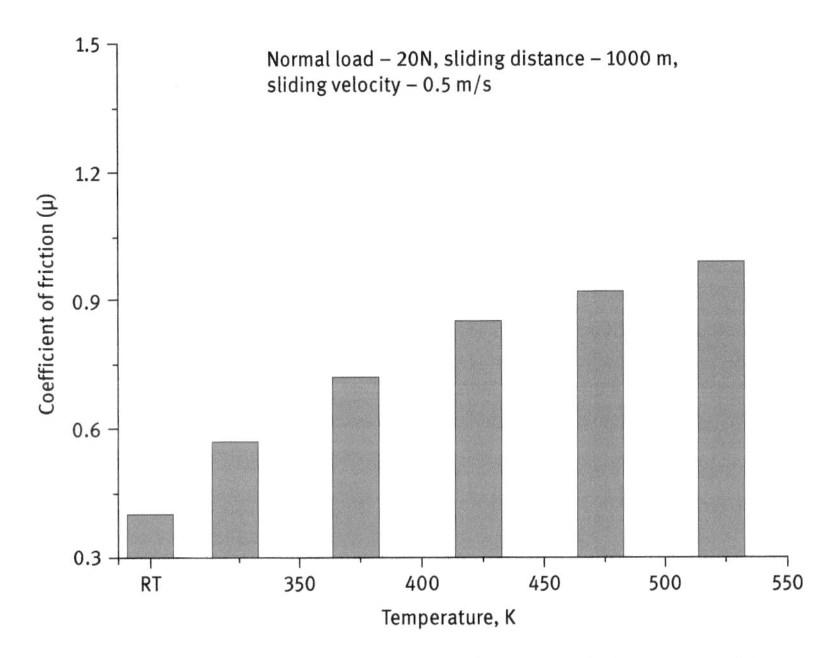

Figure 4.7: Effect of temperature on coefficient of friction of hybrid composite with constant sliding velocity, sliding distance and load.

It has been found that the type and size of the reinforcement phase in AMCs does not have much effect on the friction of composite [32]. With increasing *insitu* reaction holding time, the grain size of matrix decreases, which contributes to increase in coefficient of friction [37]. The formation of MML layer on the surface of AMCs during sliding also affects the friction and the value of COF decreases due to the lubrication nature of Fe-rich MML layer [43].

4.4 Conclusions

This chapter is an overview of the parameters affecting the tribological behaviour of AMCs with ceramic or intermetallic reinforcement.

Efforts have been made to control the losses due to wear and friction. This has been done by examining the composites for various parameters such as sliding distance, sliding velocity, applied load and temperature. Wear and friction have been thoroughly discussed for the composites having different reinforcements, and the mechanisms involved for the tribological behaviour are presented in a systematic manner. During the synthesis of composites, the effect of reaction temperature, holding time, stirring, casting process, secondary process and so on on the tribological properties has also been addressed.

References

[1] Mohan S, Mohan A. Wear, friction and prevention of tribo-surfaces by coatings/nanocoatings. In: Aliofkhazraei M, editor. Anti-abrasive nanocoatings: current and future applications. United Kingdom: Elsevier, 2015:3–22.

[2] Surappa MK. Aluminium matrix composites: challenges and opportunities. Sadhana 2003;28:319–34.

[3] Gautam G, Kumar N, Mohan A, Gautam RK, Mohan S. Strengthening mechanisms of (Al_3Zrmp+ZrB_2np)/AA5052 hybrid composites. J Compos Mater 2016;50:4123–33.

[4] Kumar N, Gautam G, Gautam, RK, Mohan A, Mohan S. A study on mechanical properties and strengthening mechanisms of AA5052/ZrB_2 in situ composites. J Eng Mater-T ASME 2017, 139, 011002-1-011002-8.

[5] Gautam G, Mohan A. Effect of ZrB_2 particles on the microstructure and mechanical properties of hybrid (ZrB_2+Al_3Zr)/AA5052 insitu composites. J Alloys Compd 2015;649:174–83.

[6] Kumar N, Singh SK, Gautam G, Padap AK, Mohan A, Mohan S. Synthesis and statistical modelling of dry sliding wear of Al 8011/6 vol.% AlB_2 insitu composite. Mater Res Express 2017;DOI:10.1088/2053-1591/aa8dbe.

[7] Singh J, Chauhan A. Overview of wear performance of aluminium matrix composites reinforced with ceramic materials under the influence of controllable variables. Ceram Int 2016;42:56–81.

[8] Suresh S, Moorthi NSV , Vettivel SC, Selvakumar N. Mechanical behavior and wear prediction of stir cast Al–TiB2 composites using response surface methodology. Mater Des 2014;59:383–96.

[9] Çelik YH, Seçilmis K. Investigation of wear behaviours of Al matrix composites reinforced with different B_4C rate produced by powder metallurgy method. Adv Powder Technol 2017;28:2218–24.

[10] Bai M, Xue Q, Ge Q. Wear of 2024 Al-Mo-SiC composites under lubrication. Wear 1996;195:100–5.

[11] Gautam G, Ghose AK, Chakrabarty I. Tensile and dry sliding wear behavior of in-situ Al_3Zr + Al_2O_3-reinforced aluminum metal matrix composites. Metall Mater Trans A 2015;46A:5952–61.

[12] Mohan S, Srivastava S. Surface behaviour of as-Cast Al–Fe intermetallic composites. Tribol Lett 2006;22:45–51.

[13] Mohan S, Gautam G, Kumar N, Gautam RK, Mohan A, Jaiswala AK. Dry sliding wear behavior of Al-SiO_2 composites. Compos Interface 2016;23:493–502.

[14] Das K, Narnaware LK. A study of microstructure and tribological behaviour of Al–4.5% Cu/Al_3Ti composites. Mater Charact 2009;60:808–16.

[15] Gautam G, Kumar N, Mohan A, Gautam RK, Mohan S. Tribology and surface topography of tri-aluminide reinforced composites. Tribol Int 2016;97:49–58.
[16] Shorowordi KM, Haseeb AS, Celis JP, Velocity effects on the wear, friction and tribochemistry of aluminium MMC sliding against phenolic brake pad. Wear 2004;256:1176–81.
[17] Ramesh CS, Ahamed A. Friction and wear behavior of cast Al 6063 based in situ metal matrix composites. Wear 2011;271:1928–39.
[18] Kumar N, Gautam G, Gautam RK, Mohan A, Mohan S. Wear, friction and profilometer studies of insitu AA5052/ZrB_2 composites. Tribol Int 2016;97:313–26.
[19] Gautam G, Mohan A. Wear and friction of AA5052-Al_3Zr in situ composites synthesized by direct melt reaction. J Tribol-T ASME 2016;138:021602-1–021602-12.
[20] Kumar N, Gautam G, Gautam RK, Mohan A, Mohan S. High-temperature tribology of AA5052/ ZrB_2 PAMCs. J Tribol-T ASME 2017;139:011601-1–021602-12.
[21] Kumar S, Sarma VS, Murty BS. Effect of temperature on the wear behaviour of Al-7Si-TiB_2 in-situ composites. Metall Mater Trans A 2009;40A:223–31.
[22] Rajan HBM, Ramabalan S, Dinaharan I, Vijay SJ. Effect of TiB_2 content and temperature on sliding wear behaviour of AA7075/TiB_2 in situ aluminium cast composites. Arch Civ Mech Eng 2014;14:72–9.
[23] Kumar S, Subramanya VS, Murty BS. High temperature wear behaviour of Al–4Cu–TiB_2 in situ composites. Wear 2010;268:1266–74.
[24] Singh J, Alpas AT. Elevated temperature wear of Al6061 and Al6061-20% Al_2O_3. Scr Metall 1995;32:1099–105.
[25] Gautam G, Kumar N, Mohan A, Gautam RK, Mohan S. High-temperature tensile and tribological behavior of hybrid (ZrB_2+Al_3Zr)/AA5052 in situ composite. Metall Mater Trans A 2016;47A:4709–20.
[26] Archard JF. Contact and rubbing of flat surface. J Appl Phys 1953;24:981–8.
[27] Kumar S, Chakraborty M, Sarma VS, Murty BS. Tensile and wear behavior of in situ Al–7Si/TiB_2 particulate composites. Wear 2008;265:134–42.
[28] Miyajima T, Iwai Y. Effects of reinforcements on sliding wear behavior of aluminum matrix composites. Wear 2003;255:606–16.
[29] Bhansali KJ, Mehrabian R. Abrasive wear of aluminum-matrix composites. J Meter 1982;9:30–4.
[30] Hosking FM, Portillo FF, Wunderlin R, Mehrabian R. Composites of aluminium alloys: fabrication and wear behavior. J Mater Sci 1982;17:477–98.
[31] Caracostas CA, Chiou WA, Fine ME, Cheng HS. Tribological properties of aluminium alloy matrix TiB_2 composite prepared by in situ processing. Metall Mater Trans A 1997;28:491–502.
[32] Roy M, Venkataraman B, Hanuprasad VV, Mahajan YR, Sundarajan G. The effect of particulate reinforcement on the sliding wear behavior of aluminium matrix composites. Metall Trans A 1992;23:2833–47.
[33] Gautam G, Kumar N, Mohan A, Gautam RK, Mohan S. Synthesis and characterization of tri-aluminide in situ composites. J Mater Sci 2016;5:8055–74.
[34] Gupta M, Srivatsan TS. Interrelationship between matrix microhardness and ultimate tensile strength of discontinuous particulate-reinforced aluminium alloy composites. Mater Lett 2001;51:255–61.
[35] Mohan A, Gautam G, Kumar N, Mohan S, Gautam RK. Synthesis and tribological properties of AA5052-base insitu composites. Compos Interfaces 2016;23:503–18.
[36] Zhao Y, Zhang S, Chen G, Cheng X. Effects of molten temperature on the morphologies of in situ Al_3Zr and ZrB_2 particles and wear properties of (Al_3Zr + ZrB_2)/Al composites. Mater Sci Eng A 2007;457:156–61.

[37] Mallikarjuna C, Shashidhara SM, Mallik US, Parashivamurthy KI. Grain refinement and wear properties evaluation of aluminium alloy 2014 matrix–TiB_2 in situ composites. Mater Des 2011;32:3554–9.

[38] Gaoa LL, Cheng XH. Microstructure and dry sliding wear behavior of Cu–10%Al–4%Fe alloy produced by equal channel angular extrusion. Wear 2008;265:986–91.

[39] Das K, Narnaware LK. Synthesis and characterization of Al–4.5%Cu/Al_3Ti composites: microstructure and ageing behaviors. Mater Sci Eng A 2008;497:25–30.

[40] Vijian P, Arunachalam VP. Optimization of squeeze cast parameters of LM6 aluminium alloy for surface roughness using taguchi method. J Mater Process Tech 2006;180:161–6.

[41] Li GR, Zhao YT, Wang HM, Chen G, Dai QX, Cheng XN. Fabrication and properties of in situ (Al_3Zr+Al_2O_3)p/A356 composites cast by permanent mould and squeeze casting. J Alloy Compd 2009;471:530–5.

[42] Rao RN, Das S, Mondal DP, Dixit G. Effect of heat treatment on the sliding wear behaviour of aluminium alloy (Al–Zn–Mg) hard particle composite. Tribol Int 2010;43:330–9.

[43] Siddhalingeshwar IG, Deepthi D, Chakraborty M, Mitra R. Sliding wear behavior of in situ Al–4.5Cu–5TiB_2 composite processed by single and multiple roll passes in mushy state. Wear 2011;271:748–59.

[44] Herbert MA, Maiti R, Mitra R, Chakraborty M. Wear behavior of cast and mushy state rolled Al–4.5Cu alloy and in situ Al4.5Cu–5TiB_2 composite. Wear 2008;265:1606–18.

[45] Antoniou R, Borland DW. Mild wear of Al-Si binary alloys during unlubricated sliding. Mater Sci Eng A 1987;93:57–72.

[46] Wilson S, Alpas AT. Effect of temperature on the sliding wear performance of Al alloys and Al matrix composites. Wear 1996;196:270–8.

[47] Uyyusru K, Surappa MK, Brusethaug S. Effect of reinforcement volume fraction and size distribution on the tribological behavior of Al-composite/brake pad tribocouple. Wear 2006;260:1248–55.

[48] Gul F, Acilar M. Effect of the reinforcement volume fraction on the dry sliding wear behaviour of Al–10Si/SiCp composites produced by vacuum infiltration technique. Compos Sci Technol 2004;64:1959–70.

[49] Ghazali MJ, Rainforth WM, Jones H. Dry sliding wear behaviour of some wrought, rapidly solidified powder metallurgy aluminium alloys. Wear 2005;259:490–500.

[50] Ravikiran A, Surappa MA. Effect of Sliding speed on wear behavior of A356 Al-30% SiC_p MMC. Wear 1997;206:33–8.

[51] Riahi AR, Alpas AT. The role of tribo-layers on the sliding wear behavior of graphitic aluminum matrix composites. Wear 2001;251:1396–407.

[52] Mohan S, Gautam RK, Mohan A. Tribology and aluminium matrix composites. In: Tyagi R, editor. Processing techniques and tribological behavior of composite materials, IGI Global, 2015, 126–48.

[53] Mandal A, Murty BS, Chakraborty M. Sliding wear behavior of T6 treated A356–TiB_2 in situ composites. Wear 2009;266:865–72.

[54] Ravikiran A, Surappa MK. Oscillations in coefficient of friction during dry sliding of A356 Al–30 wt.% SiCp MMC against steel. Scr Mater 1997;36:95–8.

[55] Rana F, Stefanescu DM. Friction properties of Al-1.5%Mg/SiC particulate metal-matrix composites. Metall Mater Trans A 1989;20:1564–6.

[56] Zhu H, Jar C, Song J, Zhao J, Li J, Xie Z. High temperature dry sliding friction and wear behaviour of aluminium matrix composites ($Al_3Zr\alpha$-Al_2O_3)/Al. Tribol Int 2012;48:78–86.

Jinjun Lu, Xiaoqin Wen, Zhiqin Ding, Feiyan Yuwen, Junhu Meng, and Ruiqing Yao

5 Silver-based self-lubricating composite for sliding electrical contact: Material design, preparation, and properties

5.1 Introduction

The tribological problem of brush-slip ring contacts to transmit power from the rotating solar arrays onto the body of the satellite is fatal to the satellite. Several silver-based self-lubricating composites (e.g. Ag-MoS_2) have been successfully used as contact materials for brush-slip ring contacts on space vehicles. Nowadays, there is an urgent demand on improving tribological property (i.e. low friction coefficient, long wear lifetime, and low electrical noise) of brush-slip ring contacts in China. Developing new silver-based self-lubricating composite is considered as one of the effective ways to improve the tribological property of brush-slip ring contacts. In this chapter, the constitutions and microstructure of the silver-based self-lubricating composite are introduced. The preparation of the silver-based self-lubricating composite, including processing method and surface finishing, is briefed. The tribological property of the silver-based self-lubricating composite and several important issues related to the tribological property are addressed. Optimization of the tribological properties based on material design and tribo-interface is briefly discussed.

Exploration of space requires developing advanced self-lubricating materials. As one of the important tribo-pairs in space mechanisms, brush-slip ring contacts transmit power from the rotating solar arrays onto the body of the satellite. The tribological problems, for example seizure, severe wear, as well as unacceptable electrical noise, on brush-slip ring contacts, are fatal to the satellite and should be avoided. Up to now, several silver-based self-lubricating composites (e.g. Ag-MoS_2) have been successfully used as contact materials for brush-slip ring contacts operating under high vacuum conditions [1, 2]. In specific, the slip rings are commonly made of coin silver, whereas the brush contacts are made of silver-based self-lubricating composites.

The new millennium sees the rapid development of space exploration of China. Space tribology is an active and challenging field for the tribologists and lubrication engineers in China. There is an urgent demand on improving tribological property (i.e. low friction coefficient, long wear lifetime, low and stable contact resistance) of brush-slip ring contacts. Developing new silver-based self-lubricating composite is considered as one of the effective ways to improve the tribological property of brush-slip ring contacts. For the last ten years, silver-based self-lubricating composite

https://doi.org/10.1515/9783110352986-005

has come a long way in China in three major Chinese institution or universities (Lanzhou Institute of Chemical Physics, Central South University and Hefei University of Technology). However, the development of silver-based self-lubricating composite for brush-slip ring contacts still lags behind the need of space exploration. Is it technically feasible to develop new silver-based self-lubricating composite of lower friction coefficient and higher wear resistance than the old ones? Or is it a time-and-money consuming effort? Therefore, it is necessary to review the state of the art of silver-based self-lubricating composite for electrical contact. There are too many papers and handbooks [1, 2] describing silver-based self-lubricating composite where readers can find detailed information. This chapter just provides a concise outline of silver-based self-lubricating composite in constitution, microstructure, preparation and tribological property. Several important issues related to the tribological property of silver-based self-lubricating composite are addressed.

5.2 Material design and preparation

5.2.1 What is silver-based self-lubricating composite?

In this chapter, silver-based self-lubricating composite refers to a free-standing part rather than coating unless otherwise stated. Silver-based self-lubricating composite is a two-phase or multiphase material with silver or silver alloy (e.g Ag-Cu) as the matrix and most commonly graphite, MoS_2 or $NbSe_2$ as solid lubricant. The word *multiphase* is used here because refractory metal (e.g. Ta, W) or one-dimensional carbon materials (e.g. carbon nanotube, carbon fiber) as a third phase is sometimes used as a strengthening component or wear resistant component but is neither the matrix nor a solid lubricant. Small amount of additive with strong electronegativity (e.g. AgCl, AgI) is believed to be used for extinguishing arc and reducing electrical noise. Table 5.1 summarizes typical silver-based self-lubricating composites for sliding electrical contact. Typical microstructure of a silver-based self-lubricating composite can be seen in Figure 5.1. For the last ten years, silver-based self-lubricating composite has come a long way in China because of great demand on exploration of space. In Table 5.1, composites (No.4 to No.8) are developed separately from three major Chinese institution or universities (Lanzhou Institute of Chemical Physics, Central South University, and Hefei University of Technology).

5.2.2 Constitutions of the composite

It is important to understand the basic constitutions (matrix, solid lubricant, a third phase) of the composite before any attempt to make good silver-based self-lubricating composite.

Table 5.1: Typical silver-based self-lubricating composites for sliding electrical contact.

No.	Composition, wt.%	Manufacturer	Friction coefficient	Wear rate
1	Ag-6.9MoS_2-2.5Cu	IMI (UK)	0.15	10^{-15} m^3/Nm
2	Ag-6.9MoS_2-2.5Cu	Le Carbone (France)	0.09	10^{-14} m^3/Nm
3	Ag-14MoS_2-3Cu-3Gr	Carbex (Sweden)	0.15	10^{-15} m^3/Nm
4	Ag-Ta-MoS_2-Gr	LICP[a] (China)	0.17	0.6 mg
5	Ag-Ta-MoS_2-AgCl-Gr	LICP (China)	0.23	1.1 mg
6	Ag-W-$NbSe_2$-Gr	LICP (China)	0.24	0.9 mg
7	Ag-6-8MoS_2-2.5Cu	CSU[b] (China)	0.15	10^{-14} m^3/Nm
8	Ag-MoS_2-Gr-CNT	HFUT[c] (China)	0.18	10^{-15} m^3/Nm

Note: [a]Lanzhou Institute of Chemical Physics, Chinese Academy of Sciences, [b]Central South University, [c]Hefei University of Technology.

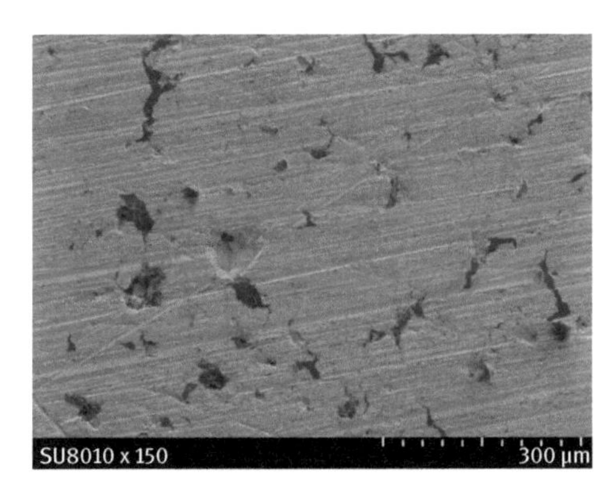

Figure 5.1: Typical microstructure of a silver-based self-lubricating composite for sliding electrical contact. MoS_2 particles are dark phase.

5.2.2.1 Matrix

As the matrix, silver is the binder for solid lubricant. Silver has good electrical conductivity and high ductility. It is prone to severe plastic deformation under high loading. Compared with pure silver, silver alloys have higher hardness and mechanical strength but lower electrical conductivity. Silver alloy (Ag-Cu alloy, see Table 5.1) is a good choice.

5.2.2.2 Solid lubricant

Typical solid lubricants of silver-based self-lubricating composite for sliding electrical contact are listed in Table 5.2. They can be classified as two groups, that is graphite and transition metal dichalcogenide (TMD, e.g. MoS_2, $NbSe_2$). Graphite is electrically conductive. For TMD, some (e.g. $NbSe_2$) are good conductor while some (e.g. MoS_2) are semiconductor. It is found that the composites has comparable electrical resistivity as pure silver when the amount of solid lubricant is less than 20%. That is why MoS_2 and

Table 5.2: Typical solid lubricants for silver-based self-lubricating composite for sliding electrical contact.

Solid lubricant	Crystal structure	Electrical resistivity, Ω×cm	Comment on tribological property
Ag	2H, Hexagonal	1.62×10^{-6}	–
Graphite	2H, Hexagonal	2.64×10^{-5}	Solid lubricant
MoS_2	2H, Hexagonal	8.51	Solid lubricant
$NbSe_2$	2H, Hexagonal	5.35×10^{-6}	Solid lubricant/ no film-forming
$MoSe_2$	2H, Hexagonal	1.86×10^{-4}	Solid lubricant

WS_2 can be solid lubricant of silver-based self-lubricating composite working under current-loading.

Graphite is a good solid lubricant for humid air or other environments where there is sufficient moisture. As an alternative to graphite as the solid lubricant, TMDs (e.g. MoS_2) are good lubricants working smoothly in vacuum or other environments where there is insufficient moisture. MoS_2 is the most frequently used solid lubricant, see Table 5.1. The main drawback of MoS_2 is its semiconducting nature. Tribo-film of MoS_2 on the worn surface not only reduces the friction and wear but also increases the contact resistance and Joule heat at tribo-interface. In contrast, $NbSe_2$ is good electrical conductor. However, there are controversial results for the tribological property of $NbSe_2$, that is good solid lubricant and no film-forming ability, see Table 5.2.

In addition to $NbSe_2$, several TMDs (e.g. $MoSe_2$) are electrical conductor and candidate solid lubricant for silver-based self-lubricating composite, see Table 5.2. Screening of TMDs for silver-based self-lubricating composite can be conducted by evaluating the tribological behavior of burnished TMDs film.

5.2.2.3 Strengthening component and wear resistant component

As seen in Table 5.1, refractory metal (e.g. Ta, W) or one-dimensional carbon materials (e.g. carbon nanotube, carbon fiber) can be used as a strengthening component or wear resistant component for silver-based self-lubricating composites.

The refractory metal (e.g. Ta, W) has high melting point and is incompatible to silver (those that form immiscible phases when in a molten state). As a result, fine Ta or W particles are uniformly distributed in silver matrix after hot pressing process. Dispersion strengthening effect of Ta or W particles increases the hardness and wear resistance of the composite without increasing the electrical resistivity of the composite [3].

Due to its excellent mechanical property, one-dimensional carbon materials (e.g. carbon nanotube, carbon fiber) can be used as a strengthening component or wear resistant component for composites. Ma reveals that the addition of short cut carbon fiber increases the wear resistance of Ag-Cu-MoS_2 composite [4]. Li reveals that the addition of carbon nanotube increases the bending strength and wear resistance of Ag-MoS_2-C composite [5].

5.2.2.4 Additive

As mentioned in Section 1.2.1, inorganic additive with strong electronegativity can be used for extinguishing arc and reducing electrical noise because it reduces the work function of sliding surface. Zheng reveals that the addition of silver halides (AgCl and AgI) reduces the wear rate of Ag-MoS_2 composite and electrical noise [6].

5.2.3 Preparation

Powder metallurgy (P/M) is the most frequently used technique for fabrication of a silver-based self-lubricating composite. Although pressureless sintering is suitable for large-scale production and hot pressing is ineffective for production, silver-based self-lubricating composites are prepared by using hot pressing in vacuum or nitrogen for higher mechanical strength. Hot extrusion is sometimes used for better mechanical property.

The preparation of a P/M silver-based self-lubricating composite starts from powders of silver or silver alloy (mixed metal or pre-alloy) and solid lubricants. The source, composition and particle size and shape of the powders are crucial for the properties of the composite.

Mixing of the starting powders is also an important step of P/M. As silver and silver alloy are of high ductility, sometimes high energy ball milling is not suggested. Sintering temperature and atmosphere can be important.

Surface finishing should be handled with great care because silver-based self-lubricating composites are soft. Abrasive particles can be embedded during surface finishing and this has a detrimental influence on the tribological property of brush-slip ring contacts.

5.2.4 Material design

The basic design of silver-based self-lubricating composite is quite simple, that is matrix for mechanical property and electrical conductivity while solid lubricant for self-lubricating characteristic. In practice, such a composite is too simple to meet the requirement. In the following subsections, material design based on three important aspects, that is strategy for material design, interface in the composite and functionally graded material, is presented.

5.2.4.1 Strategy for material design

The material design for silver-based self-lubricating composite is a property-based design. Mechanical strength (e.g. hardness), electrical resistivity, friction and wear, contact resistance are the main properties of a silver-based self-lubricating composite for electrical contact. It is obvious that different strategies are required for the four properties. For example, solid solution strengthening and dispersion strengthening increase the mechanical strength. However, solid solution strengthening increases the electrical resistivity of the composite. High amount of solid lubricant reduces friction and wear but decreases the mechanical strength of the composite. MoS_2 film at tribo-interface reduces friction and wear but increases the contact resistance. For low contact resistance, a metal to metal contact is preferable however; it results high friction and severe wear. It is evident that the some strategies shown in Table 5.3 are contradictory. In this sense, it is impossible to fabricate silver-based self-lubricating composite with combined properties of high mechanical strength, low electrical resistivity, low friction and wear, low contact resistance.

Table 5.3: Some strategies for high mechanical strength, low electrical resistivity, low friction and wear, low contact resistance and their interrelation.

Strategy	Mechanical strength	Electrical resistivity	Friction and wear	Contact resistance
Solid solution strengthening	+	–	N/A	N/A
Dispersion strengthening	+	±	+	N/A
Solid lubricant	–	±	+	±
Metal to metal contact	N/A	N/A	–	+

Note: + means positive effect, – means negative effect, ± means positive effect or negative effect. N/A means not available.

5.2.4.2 Interface in the composite

For a silver-based self-lubricating composite, the interface between matrix and solid lubricant plays an important role in determining the tribological and mechanical properties. We know very little about the role of the interface in sliding contact. There are too many questions on the interface. For example, should it be weak or strong? There is still no a definite answer. By using silver-coated MoS_2 powders, the composite has higher mechanical strength and wear resistance than that using raw MoS_2 powders.

5.2.4.3 Functionally graded material

As mentioned in Section 5.2.4.1, it is impossible to fabricate silver-based self-lubricating composite with combined properties of high mechanical strength, low electrical resistivity, low friction and wear, low contact resistance. Zhang proposes a design based on idea of functionally graded material [7]. The functionally graded composite is composed of three layers, that is a working layer, a transition layer, and a welding layer. This structure renders the composite good electrical property, tribological property, and welding performance.

5.3 Tribological properties

5.3.1 General view

The electrical sliding properties required for a contact material for brush-slip ring contacts in space application are as follows. (1) Low friction coefficient and ability to reduce frictional heat and eliminate adhesion. (2) High wear resistance and less wear debris as possible. (3) Low contact resistance. (4) Low electrical noise. (5) Good tribological properties after long-term storage. The operating conditions of brush-slip ring contacts are diversity in sliding speed (from 0.1 rpm to several hundred rpm) and current density. The detailed information on the tribological property of silver-based self-lubricating composite as brush can be found in Handbook of Space Tribology [2]. The service lifetime of brush-slip ring contacts can be as long as 10 years or even longer. In addition, there is an increasing demand for low contact resistance and low electrical noise. As such, understanding the important issues of the tribological property of silver-based self-lubricating composite as brush is of importance.

5.3.2 Three important issues

We focus on the three important issues related to the tribological property of silver-based self-lubricating composite, that is unlubricated or lubricated, running in and genesis of wear debris, and geometry effect of brush and slip ring. The three issues are definitely not all the important issues for the tribological property of silver-based self-lubricating composite. However, the authors value the importance of the three issues.

5.3.2.1 Unlubricated or lubricated?

Unlubricated or lubricated? It is a question. Comparison on unlubricated and lubricated brush-slip ring contacts using silver-based self-lubricating composite as brush can be dated back to 1970s [2]. The conclusion is that the performance of the unlubricated is better than that of the lubricated. For example, brush-slip ring lubricated by low volatility oil works smoothly for four months and suffers severe wear and large electrical noise afterwards. This is due to temperature rise at the tribo-interface of brush and slip. As a result, loss of oil lubricant leads to severe wear and large electrical noise.

Room-temperature ionic liquids (ILs) are molten salts, which are liquid at room temperature and that generally consist of bulky organic cations paired with organic or inorganic anions. ILs present a unique combination of properties, for example nonflammability, high-thermal stability, wide liquid range, and, most relevant, negligible vapor pressure. They are also good lubricant and have advantages over conventional lubricating oil. The use of ILs might open a new venue to lubricated brush-slip ring contacts.

5.3.2.2 Running in and genesis of wear debris

Running in is very important for silver-based self-lubricating composite. It is suggested that running in should be necessary for a long time before stable contact between brush and slip ring. Because the wear rate of the brush during running in is one order of magnitude higher than that in stable stage, a large number of flake-like wear debris are generated. Understanding the running in process and genesis of wear debris is very helpful. A question "Is it possible to develop a *fast* running in process of *minimum damage* to brush?" deserves a great effort to answer. The wear lifetime and variation of contact resistance of brush-slip ring can be simulated by accelerating experiment, which is time-saving. Tribologists should be able to accelerate the running in process and minimize the wear and plastic deformation of the brush.

5.3.2.3 Geometry effect of brush and slip ring

The brush has a smaller area while slip ring has a larger area in sliding contact, which is the basic configuration for brush-slip ring contact. "Rider wear" by Antler [8] occurs to the brush made of silver-based self-lubricating composite. Geometry effect of brush and slip ring should be considered minimizing the wear of brush.

5.3.3 Outlook

Now we face the bottleneck for developing silver-based self-lubricating composite and call for new strategies. For material aspect, two-dimensional materials (e.g. graphene, few layer TMDs) can be used as solid lubricant for better tribological property and mechanical property as well. Functionally graded material also provides an effective way for combined electrical property, tribological property and welding performance on a brush. For tribology aspect, understanding the running in process and genesis of wear debris is very helpful. Surface finishing of the composite can be important as well. Tribologists must bear in mind that design of tribo-interface between brush and slip ring is the most important issue.

Acknowledgment

The authors extend their grateful thanks for the financial support from Natural Science Foundation of China (No. 51775434) and Opening Project Foundation of State Key Laboratory of Solid Lubrication (LSL-1301).

References

[1] Masahisa M. Handbook of solid lubrication (in Chinese). Beijing: Mechanical Industry Publishing Co. Pte. Ltd., 1986.
[2] Roberts EW. Handbook of space tribology. RDl, Riley, Warrington, Cheshbv, WA3 6AT, UK, AEA Technology Ple, 2002.
[3] Zheng J, Ouyang J, Zhu J. Study on long life electrical brush slip-ring in vacuum (in Chinese). J Tribol 1997;17:129–39.
[4] Ma C, Wang X, Zhang L, Zhou K. Friction and wear properties of Ag-MoS_2/carbon fibers reinforced composites (in Chinese). Chin J Nonferrous Met 2012;22:3074–80.
[5] Li S, Feng Y, Chen S. Effect of electric current intensity on friction and wear properties of CNTs-Ag-MoS_2-G composites in electrical sliding contact. J Mater Eng 2008;1:45–50.
[6] Zheng J, Zhu J, Ouyang J. Effects of silver halides on the properties of electric contact materials (in Chinese). J Tribol 1989;9:164–8.
[7] Zhang L, Zhou K, Liu W, Zhou R, Li S. Preparation and properties of Ag-Cu-MoS_2 brush materials (in Chinese). Chin J Nonferrous Met 2005;15:1766–9.
[8] Antler M. Wear of electroplated gold. ASLE Trans 1962;11:240–60.

Satheesan Bobby, Mohammed Mehdhar Al-Mehdhar,
and Mohammed Abdul Samad

6 Tribology of CrC–NiCr cermet coatings

6.1 Introduction

Since its inception, thermal spray technology has been an *in-vogue* field of research within academic and industrial circles. Extensive studies on thermal spray processes and advanced materials within this technological domain are available in literature; however, a significant gap seems to exist when it comes to a comprehensive literature review of developments within the last decade with emphasis on chromium carbide–nickel chromium (CrC–NiCr) hard metal coatings as a potential cost-saver for the oil and gas, aerospace and power industries. In this chapter, an outline of four broad thermal spray techniques, namely combustion, electric wire-arc, plasma and laser deposition, has been presented followed by a focused section on production of CrC–NiCr coatings through various applicable routes. An in-depth critical review of various tribological tests carried out in recent years on CrC–NiCr coatings under a wide range of conditions forms the crux of this work. Performance enhancement techniques used by researchers in recent years have also been highlighted. While recent advancements in tribological characterization of CrC–NiCr coating systems have been critically examined, new directions for research have also been proposed.

The present economic climate has fueled the need to develop low-cost alternatives to conventional maintenance practices. Thermal spraying, a popular coating technique, has found its use in the oil and gas, marine, aerospace, automotive and power sectors to name a few [1]. A major part of the maintenance spending of oil and gas companies comprises the selection and application of coating systems to protect critical assets from the adverse effects of corrosion and other wear modes over the life cycle of the equipment. An extensive report on application areas and potential opportunities in the oil and gas industry with respect to thermally sprayed hard metal coatings has been compiled by Mcdonald [2]. Coating thicknesses ranging from a few micrometers to several millimeters can be obtained using this technique. This technique can also be used to repair damaged parts and restore repair dimensions. The major focus of various studies in the oil and gas segment has been the protection of rotor journal areas (journal bearings that support the rotor/shaft of turbines) and blades from wear and this has undoubtedly extended the reliability and life cycle of such critical equipment [3, 4]. It is also one of the most popularly employed repair and protection methods for multiwear mode environments such as turbine and compressor parts [5–7]. The spectrum of materials that can be thermally sprayed is practically huge and range from metals to alloys to ceramics and cermets.

https://doi.org/10.1515/9783110352986-006

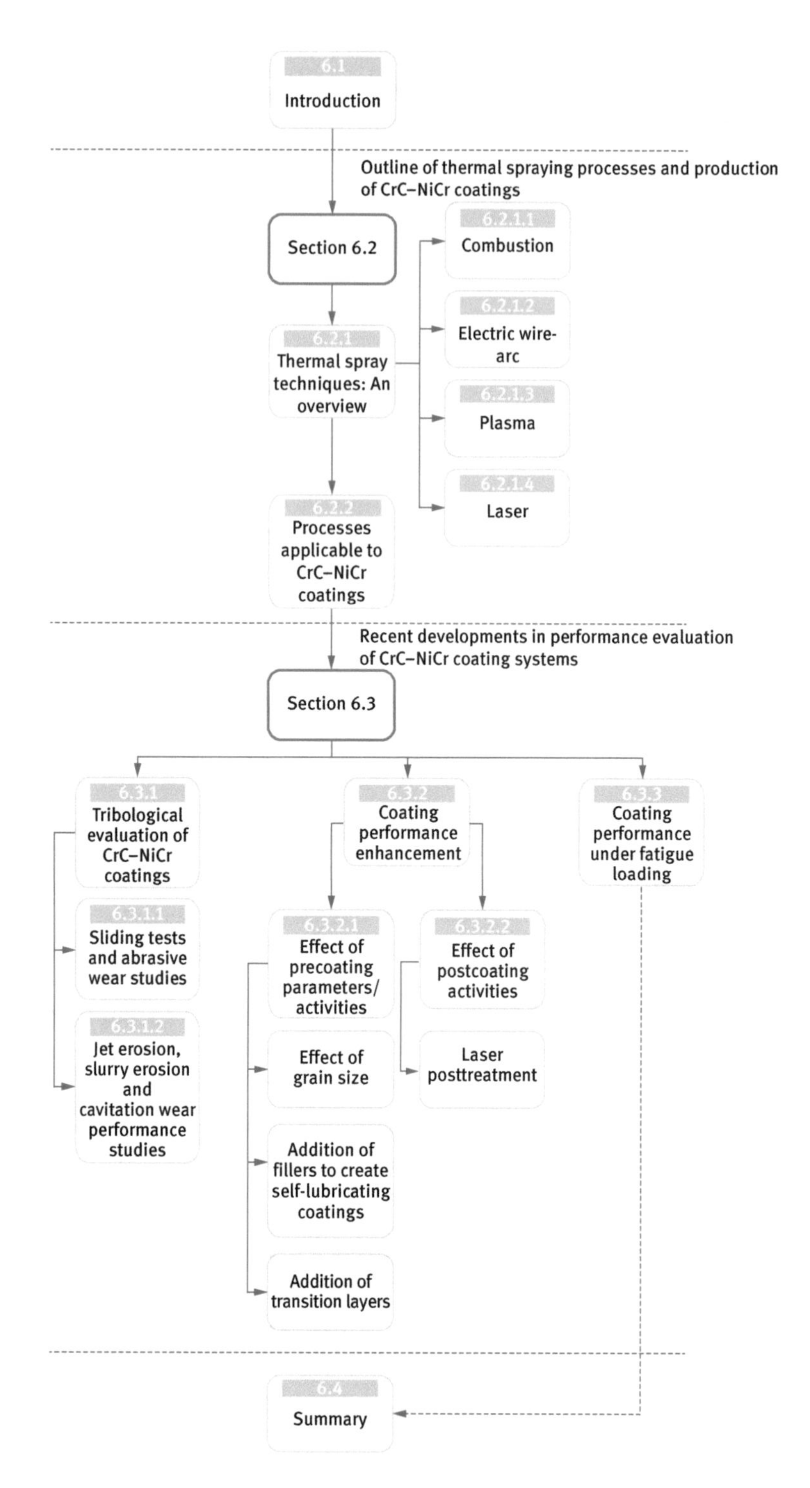

Figure 6.1: Flowchart showing the high-level structure of the chapter.

A cermet is a composite material comprising metal(s) and ceramic(s) in various ratios to obtain a material with enhanced properties including, but not limited to, corrosion resistance, wear resistance and thermal stability. A wide range of cermets have been synthesized over the years and their selection as a coating system for an application is governed largely by the operating parameters of the equipment and combination of wear modes experienced by the substrate in question. The most common of cermet coatings, possessing excellent microstructural and mechanical properties, used with excellent results on a range of applications are CrC–NiCr, WC–Co (tungsten carbide–cobalt) and WC–CoCr (tungsten carbide–cobalt chromium) [8]. Although WC–Co and WC–CoCr coatings show higher microhardness and wear resistance as compared to CrC–NiCr coatings [9–12], the acceptably high hardness, excellent wear resistance and better elevated temperature corrosion resistance [13] of CrC–NiCr coating systems definitely make this an attractive option to the oil and gas, aerospace and powder engineering industries. In addition to the above, CrC–NiCr coatings are also lighter and can sustain higher service temperatures of around 850 °C [14–16]. It is for these reasons that the authors have focused this comprehensive review of works carried out in recent years on chromium carbide–nickel chromium (CrC–NiCr) hard metal coatings. The flowchart presented in Figure 6.1 is intended to provide the reader with an overview of the contents and organization of this chapter.

6.2 Outline of thermal spraying processes and production of CrC–NiCr coatings

In this section, different thermal spray techniques have been outlined followed by a focused subsection on production of CrC–NiCr coatings through one or more of the several routes described. A summary of the CrC–NiCr coating microstructures produced through these various routes has also been presented in this section.

6.2.1 Various thermal spray techniques: An overview

6.2.1.1 Combustion

The first trial of a thermal spray coating was in the early twentieth century by Schoop using combustion technique to generate a heat source [17]. This technique has been widely used in both research and industry and has been widely employed to spray apply CrC–NiCr coatings [18, 19]. To add on, the most popularly employed combustion technique is high-velocity oxy-fuel (HVOF) which uses a combination of thermal and kinetic energy transfer to melt the coating material (in powder or wire form) and propel it onto the substrate [20]. Supersonic speed jets of molten powder are achieved by the high jet temperatures produced during the combustion process. Earlier works have suggested maintaining an optimum stoichiometric ratio between the fuel and oxygen to avoid coating contamination via soot formation or oxidation based on the richness or leanness of the mix [21]. Several modeling works on HVOF systems, albeit not related to CrC–NiCr coatings, have also been carried out by various authors [21–23]. Ma et al. [24] developed NiCrBSi coating on an AISI 316L substrate using HVOF and HVOLF (where LF stands for liquid fuel). It was found that the HVOLF process (wherein powder mixed in a liquid solvent is pumped to the combustion chamber using a modified liquid injector) produced a better performing NiCrBSi coating with less oxygen percentage (less oxidization), higher hardness, higher ductility and higher adhesive and cohesive strengths. This was attributed to the nanograined microstructure produced by the HVOLF method. A similar technique that developed with time is the high-velocity air fuel (HVAF) process and replaced oxygen with air. Since air contains relatively less oxygen when compared to a 100% saturated oxygen system, the heat generated in the combustion chamber for the same volume of gas (air or oxygen) will be less in the HVAF system. This will result in the powder particles to be heated below their melting point but accelerated to a higher velocity (above 700 m/s) to form dense and non-oxidized deposits with minimal thermal deterioration and excellent efficiency. This technique is also referred to as "warm kinetic spraying" because of the moderate thermal energy generated during the process. Furthermore, the gas temperatures developed are much lower, which prevent the particles from being superheated and helps preserve the nanocrystalline/amorphous structure of the deposits [21]. Depending on the material and process parameters, HVAF can prove to be a better process for coating when compared to HVOF specifically from a degree of oxidation point of view [25].

6.2.1.2 Electric wire-arc

The first trial of a thermal spray coating where electric arc technique was used as a heat source was also developed by Schoop after his invention of the combustion thermal spray process [17]. In this process, the coating material is in the form of an electrode where the arc strikes the material to melt it down. Gas (usually inert) is forced to flow at high pressure and speed through the molten material to atomize and propel it to the substrate surface. Gas velocities are in the order of a few hundreds of meters/second but the gas is practically not heated by the arc, allowing one to keep the substrate temperature below a few tens of degrees centigrade without cooling. Particles in wire-arc process are generally heated to temperatures above the melting temperature (~4,000 °C) and accelerated to a velocity of 100–130 m/s [26].

6.2.1.3 Plasma

In a recent study by Zhang [27], plasma spraying (PS) was stated to be the most versatile (and sophisticated) of all known thermal spraying methods. It was also reported to render a dense coating with low porosity owing to high plasma temperatures. In this process, ionization of the gas/air medium generates sufficient thermal energy to melt a wide range of materials. Hence, for most of the applications requiring hard ceramic coatings, PS technology is used. There are different types of PS methods; the most common ones being (a) atmospheric plasma spray (APS), (b) vacuum plasma spray (VPS) and (c) low-pressure plasma spray (LPPS). PS techniques in general ought to be the optimum choice when low particle velocities are required at very high temperatures. APS can perform well at 1,200 °C producing particles at velocity of 150–400 m/s. If higher traveling velocity is required, VPS/LPPS can be employed to generate particle accelerations of up to 600 m/s [26]. PS technique however has some limitations, which make it a good option for low-stress applications only as may be noted in the latter part of this chapter.

6.2.1.4 Laser

Using laser in thermal coating operations has been usually associated with posttreatment (e.g., peening) rather than being a prime process for depositing the coating particles [13]. Vuoristo et al. [28] studied the laser coating process (also referred to as laser spraying or laser cladding) and illustrated the differences between this technique and other thermal spray technologies. It was reported that the laser coating process helps produce a metallurgical bond of the coating to the substrate and produce coatings with uniform composition and thicknesses. Venkatesh et al. [29] studied the microstructural properties and phase changes occurring in laser clad CrC–NiCrMoNb coatings. An interesting finding from their work was the observance of Cr_7C_3 (rendered to be a stable phase) and Ni phases in the clad layers though starting powders contained both Cr_3C_2 and Cr_7C_3 phases. As expected, the hardness of the clad layers was found to improve with the carbide content. As opposed to earlier works, the authors employed electron backscatter diffraction (EBSD) analyses to accurately determine the amount of carbide present in the starting powders, laser clad layers and further to heat treatment.

6.2.1.5 Hybrid spray techniques

From the several thermal spray techniques presented, it can be deduced that among other factors, the properties of the end-product also depend on the spray technique employed. An interesting approach would be to think of combining two or more techniques in one spray system and observe the results. Stanisic et al. [30] studied the arc/HVOF hybrid process to explore the physics involved and address qualitatively and quantitatively the ability of this system to atomize and deposit materials effectively. The experiments were set up under three modes for comparison purposes: (a) two-wire arcing only; (b) full hybrid using two wire arcing + powder through the HVOF feed line; and (c) HVOF mode only with material fed in powder and wire forms. It was found that the arc-HVOF separation distance had a significant effect on particle velocity. Unfortunately, the study did not show any significant improvement or superiority of the hybrid system over the conventional ones in terms of coating microstructure or mechanical properties.

It can be safely concluded that the HVOF thermal spray process continues to be widely employed and is popular in both academia and industry owing to its versatility and performance under a range of conditions. Table 6.1 presents a general overview of the HVOF process and its superiority over the PS technique for reference purposes.

Table 6.1: Comparison between HVOF and plasma spray technique.

Application	Spray process	Preference	Concerns	Remedy/improvement strategy	Negative impacts from the remedy
Fatigue resistance	HVOF	Preferred	–	Grind the coating to achieve smoother finish and improved RCF performance	–
	Plasma	Not preferred	Insufficient densification of the coating layer owing to low impact velocity of particles	–	–
Corrosion resistance	HVOF	Preferred	Splashing might occur if particle velocity is increased	Reduce the stand-off distance to shorten in-flight time Add a curvature to the substrate to reduce splashing	Reducing the stand-off distance might not always lead to satisfactory results
	Plasma	Not preferred	Existence of discontinuities leading to cracks	Annealing	Drop in hardness
Thermal cycling	HVOF	Preferred	–	–	–
	Plasma	Not preferred	Existence of discontinuities leading to cracks	Annealing	Drop in hardness

Note: HVOF, high-velocity oxy-fuel; RCF, rolling contact fatigue.

6.2.2 Thermal spray techniques associated with CrC–NiCr coatings

Of the various thermal spray techniques detailed in Section 6.2.1, not all routes have been employed to produce CrC–NiCr coatings. Combustion remains by far the most popular technique used for thermally spraying CrC–NiCr coatings with the HVOF version being more commonly adopted than the HVAF method. Other techniques such as PS and laser deposition have also been used by some researchers but to a much lesser extent. Our review of literature did not reveal works involving production of CrC–NiCr coatings using the electric wire-arc method. The aim of this section is to provide the reader with an overview of the various routes used to produce CrC–NiCr coatings along with the process parameters used and observed differences in microstructural properties. Both Tables 6.2 and 6.3 summarize the authors' findings from this chapter. While Table 6.2 summarizes the spray parameters used by different researchers and the corresponding microstructural characteristics such as hardness, porosity and oxide/carbide contents, Table 6.3 presents the scanning electron microscope (SEM) scans of the powder morphologies/coatings produced along with a summary of predominant factors determined in the corresponding study. Hence, Tables 6.2 and 6.3 need to be read in conjunction with each other.

The findings presented in Tables 6.2 and 6.3 reveal that microstructural characteristics go a long way in affecting the performance of the coating system due to its influence on physical properties, tribological and fatigue performance. Furthermore, it must be noted that all microstructures presented belong to as-sprayed coatings only. There are several techniques used to enhance the properties of the coating through microstructural modifications. Performance tests carried out on *as-sprayed* and *modified* coatings will be explored in the following sections:

6.3 Recent developments in performance evaluation of CrC–NiCr coating systems

Having seen the effect of various spray processes on the microstructure in the previous section, the performance evaluation of the produced coating systems logically follows. For clarity, this section has been subdivided into three parts (Sections 6.3.1, 6.3.2 and 6.3.3) as outlined in Figure 6.1.

6.3.1 Tribological evaluation of CrC–NiCr coatings

The increasing requirements of production processes and needs to reduce costs have driven a significant amount of interest in understanding the friction and wear characteristics of coating systems in a variety of tribo-environments. The performance

Table 6.2: Summary of spray techniques used to produce CrC–NiCr coating systems and observed microstructural characteristics.

Ref.	Route	Substrate	Thickness	Spray parameters				Microstructural characteristics		
				Gun type	Fuel	Powder feed rate	Powder type or size	Hardness	Porosity[a]	Carbide/oxide content
[6]	HVOF	Steel	150 µm	C-CJS with k5.2 nozzle	Kerosene	4 kg/h	 10–30 µm <10 µm (fine)	Method: Universal hardness, HU (under 100 mN load) 7.75 GPa 6.71 GPa	ND	ND
[13]	HVOF	Steel	350 µm	K2 gun	Paraffin	6.6 kg/h	Woka 7303 11–40 µm	Method: Lee and Gurland model $HV_{0.1}$ ~1,100	ND	ND
[12]	HVOF	ASTM-SA213-T22 steel	330 µm	HIPOJET-2100	LPG	1.8 kg/h	15–45 µm	HV = 850	2.5–3.5%	ND
[14]	HVOF	Mild steel	400 µm	GMA Microjet	Propane	0.7 kg/h	Woka 2075	$HV_{0.3}$ = 1,036	ND	Carbide content: 36%
[31]	HVAF	Mild steel	400 µm	Aerospray 150	Kerosene	2.1 kg/h	Woka 2075	$HV_{0.3}$ = 1,163	ND	Carbide content: 67%
[32]	Plasma	Mild steel	300 µm–400 µm	Hypersonic PS gun	NA	1.8 kg/h	5–70 µm	ND	ND	ND
[29]	Laser	ASTM SA 516 steel	ND	6 kW diode laser type LDF 6000	NA	0.84 kg/h	5–30 µm CrC–NiCrMoNb	Methodology: Diamond penetration; values presented at the surface $HV_{0.1}$ = 1,250 (laser power 1,600 W) $HV_{0.1}$ ~900 (laser power 2,000 W)	ND	ND

Note: HVAF, high-velocity air fuel; HVOF, high-velocity oxy-fuel; CJS, carbide jet spray.
[a]ND, not determined in the referenced work.

Table 6.3: SEM images and summary of factors influencing properties of as-sprayed coatings.

<table>
<tr><th>Ref.</th><th>SEM images</th><th>Factors influencing mechanical, tribological or fatigue properties</th></tr>
<tr><td>[6]</td><td>

Figure 6.2a: Standard powder morphology.
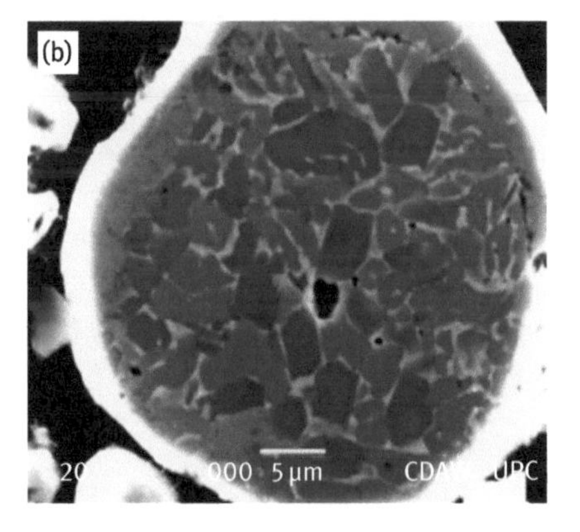

Figure 6.2b: Fine powder morphology.</td><td>1. Carbide morphology (rounded is better)
2. Carbide size (fine size is considered better)
3. Applicable to pin-on-disk (sliding wear) tests carried out under dry and lubricated conditions</td></tr>
<tr><td>[13]</td><td>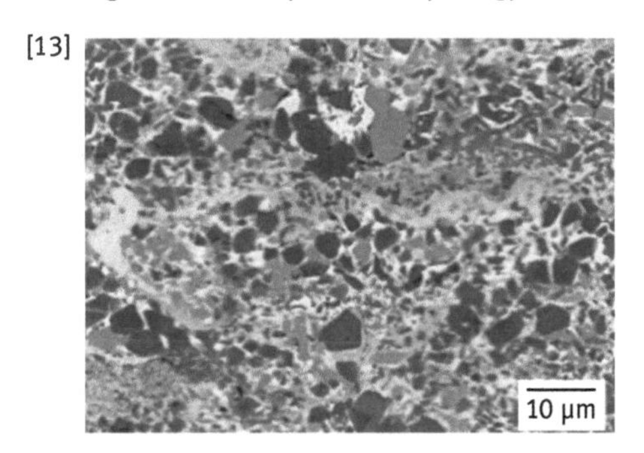

Figure 6.3: BSE image of the microstructure of as-sprayed CrC–25NiCr coating with dark areas representing carbide phases and light areas representing the binder matrix.</td><td>1. Authors regard hardness as only a secondary source of information to characterize wear performance
2. Under low-stress abrasion, the bare-hard phases were postulated to carry the load
3. Fracture toughness was considered to be the dominant factor influencing tribological performance especially under high stress abrasive wear regimes.</td></tr>
</table>

Table 6.3: (continued)

Ref.	SEM images	Factors influencing mechanical, tribological or fatigue properties
[12]	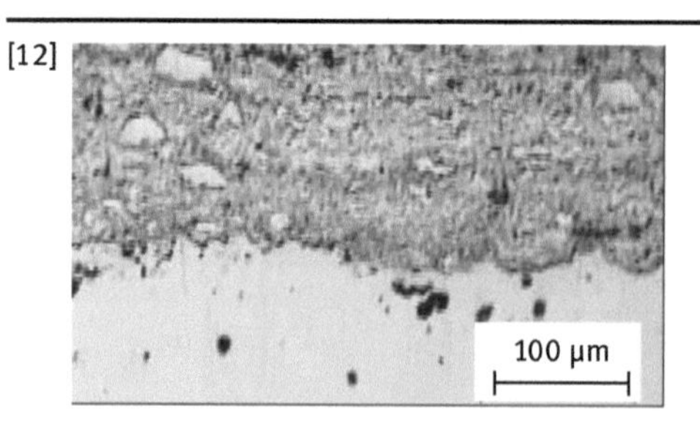 **Figure 6.4**: Optical micrograph of the as-sprayed CrC–NiCr coating.	1. Comparisons were drawn to WC-Co and hardness was postulated to influence **adhesive wear resistance** under sliding conditions 2. Formation of oxide layer that prevents metal-to-metal contact and reduces wear
[14] [31]	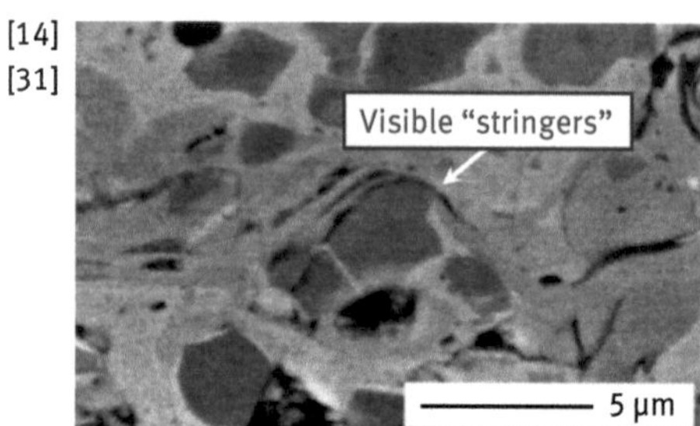 **Figure 6.5**: Stringers in the as-sprayed HVOF coating representing the oxide phases (BSE imaging). 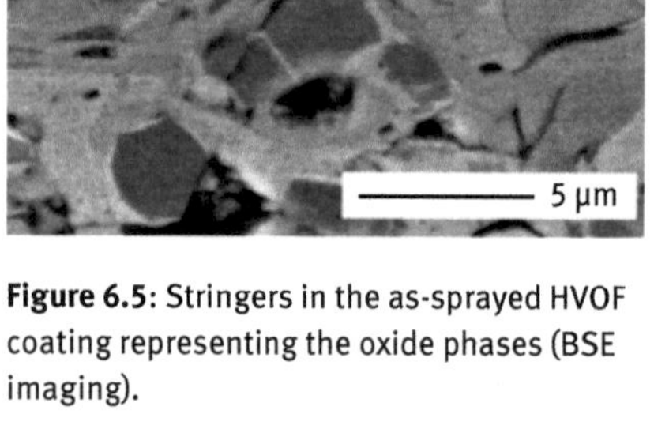 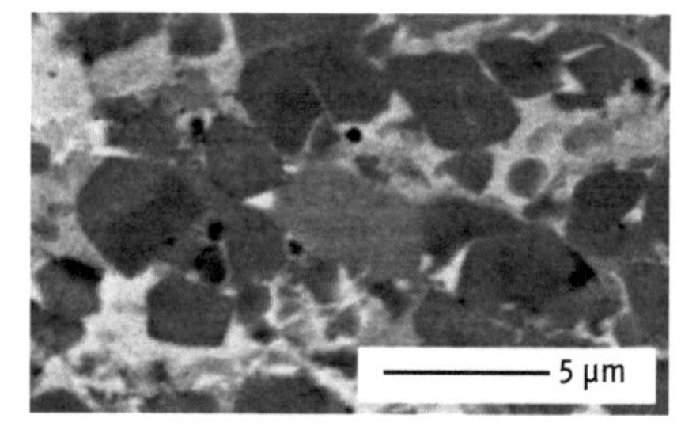 **Figure 6.6**: BSE image of as-sprayed HVAF coating.	1. High inter-splat adhesion strengths were observed to be the primary factor affecting the **erosion wear** resistance.

Table 6.3: (continued)

Ref.	SEM images	Factors influencing mechanical, tribological or fatigue properties
[32]	(a) **Figure 6.7a:** Morphology of the CrC–NiCr powder. 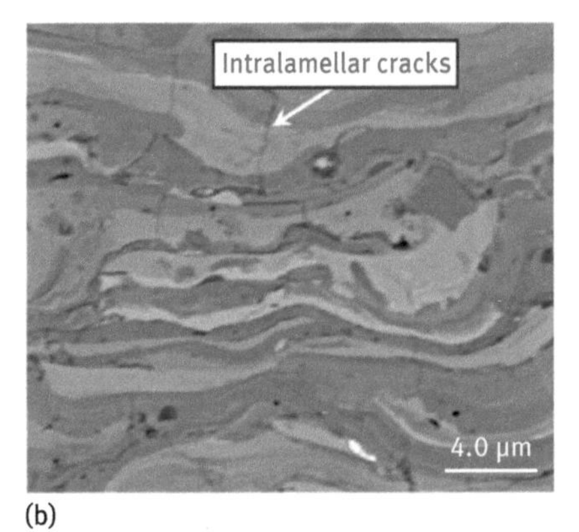 (b) **Figure 6.7b:** Cross-sectional microstructure of the as-sprayed coating showing intralamellar cracks arising from high residual stresses produced during the solidification process.	1. Coating microstructure (excessive microcracks within the lamellae are detrimental to *fatigue life* performance) 2. Residual stresses (compressive stresses are better). In this regard, HVOF method is considered better than PS.

Table 6.3: (continued)

Ref.	SEM images	Factors influencing mechanical, tribological or fatigue properties
[29]	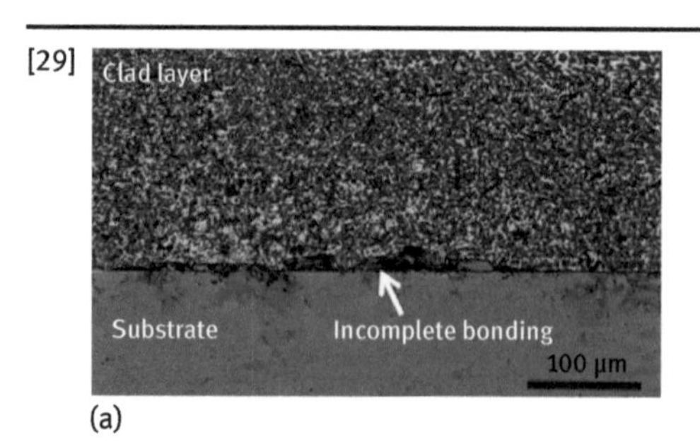 (a) **Figure 6.8a**: BSE image of laser clad coating at 1,400 W laser power and 8 mm/s scan speed. 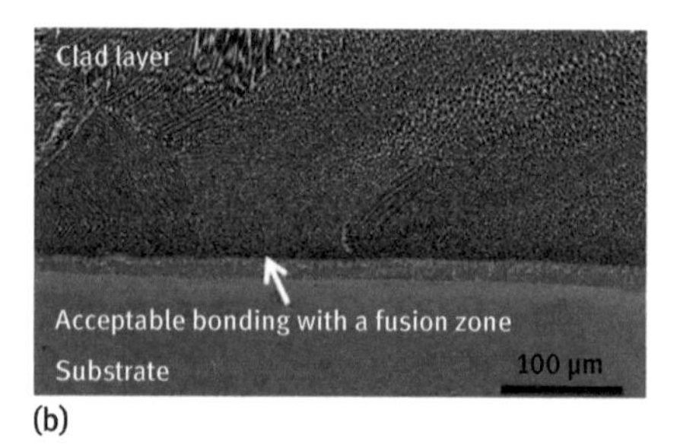 (b) **Figure 6.8b**: BSE image of laser clad coating at 1,600 W laser power and 8 mm/s scan speed.	1. Laser power ought to be greater than or equal to 1,600 W to ensure good bonding/fusion with the substrate 2. Increasing laser power increases dilution but no such effect of scan speed on dilution 3. Increasing scan speed at a given laser power reduces carbide content but no effect on **hardness**

of a coating in terms of friction coefficient or wear resistance is specific to the selected test environment and cannot be generalized. Furthermore, and since both friction and wear are system properties as opposed to material properties, it is important to correlate the coating microstructure with the test environment (loads, speeds, temperature and lubrication). This fundamental aspect of tribology in which the property of a system of components is determined is what makes a study quite challenging. As such, it may be noted that one test alone cannot completely characterize the tribological behavior of a coating system. However, tests carried out under similar conditions can be used as a screening mechanism to select the best coating system for a given set of environmental parameters. Researchers in recent years have used both customized and standardized methods to evaluate failure mechanisms and performance of coatings under different scenarios. This section includes recent tribological studies carried out on the CrC–NiCr coating system, and efforts have been made to link findings from various studies. Evaluations carried out against international standards, namely, ASTM G65 (Standard Test Method for Measuring Abrasion Using the Dry Sand/Rubber Wheel Apparatus), ASTM G76 (Standard Test Method for Conducting Erosion Tests by Solid Particle Impingement Using Gas Jets), ASTM G77 (Standard Test Method for Ranking Resistance of Materials to Sliding Wear Using Block-on-Ring Wear Test), ASTM G99 (Standard Test Method for Wear Testing with a Pin-on-Disk Apparatus), ASTM F1978 (Standard Test Method for Measuring Abrasion Resistance of Metallic Thermal Spray Coatings by Using the Taber Abraser) and ASTM G32 (Standard Test Method for Cavitation Erosion Using Vibratory Apparatus) have also been discussed. Where possible, comparisons have also been drawn to WC-based cermet coatings. Table 6.4 presents a summary of only those works carried out against international standards.

Table 6.4: Standardized tests carried out on cermet coatings in recent years.

Ref.	Coating	Technique	Test standard					
			ASTM G65	ASTM G76	ASTM G77	ASTM G99	ASTM G32	ASTM F1978
[6]	CrC–25(NiCr)	HVOF				✓		
[12]	CrC–NiCr WC–Co	HVOF				✓		
[13]	CrC–25(NiCr)	HVOF	✓					
[14]	CrC–25(NiCr)	HVOF, HVAF		✓				
[33]	CrC-25(NiCr) CrC–40(NiCr)	HVOF				✓		
[34]	CrC–20(NiCr)	HVOF	✓			✓		
[35]	CrC–25(NiCr)	Plasma			✓			
[36]	WC–10Co4Cr	HVOF, HVAF	✓			✓		
[37]	CrC–25(NiCr)	HVOF, HVAF	✓			✓		
[38]	WC–12Co	HVOF		✓				
[39]	CrC–20(NiCr)	HVOF		✓				
[40]	WC–12Co WC–10Co4Cr CrC–25(NiCr)	HVOF						✓
[41]	WC–12Co WC–12Co-50(NiCr) WC–12Co/NiCrAl interlayer	HVOF Plasma					✓	

HVOF, high-velocity oxy-fuel; HVAF, high-velocity air fuel.

6.3.1.1 Sliding tests and abrasive wear studies

The loading regimes and wear mechanisms associated with sliding wear tests such as those seen in pin-on-disk tests, ball-on-disk tests and ring-on-disk tests are quite different from the kind experienced under abrasion wear tests such as dry-sand/falling sand abrasion tests. However, both these tests have been included under this section.

Through a series of recent works, Bolelli et al. [36, 37] extensively compared the tribological behavior of HVOF and HVAF-sprayed WC-10Co4Cr and Cr_3C_2-25NiCr coating systems under similar conditions. In the first of the series of these studies [36], works were focused on evaluating the performance of WC-10Co4Cr coating under dry sliding (at room and elevated temperatures) and abrasive wear conditions. Powders of two different particle size distributions (fine ~30 μm, coarse ~45 μm) were sprayed onto a low carbon steel substrate using four different techniques (two different types of HVOF and HVAF processes) leading to a total of eight possible combinations. The microstructural characteristics of the generated coatings have been summarized in Table 6.5.

The resulting coating thickness in all cases was noted to be between 250 and 300 μm. Sliding wear tests were carried out using a ball-on-disk setup in accordance with ASTM G99 using sintered alumina balls of 6 mm diameter as the counterpart. Tests were performed under a normal load of 10 N at a sliding velocity of 0.1 m/s with a wear track radius of 7 mm across a sliding distance of 5,000 m. The sliding wear under room temperatures for the ASTM G99 test setup was observed to be minimal. However, and upon closer analysis, it was revealed that the HVOF-sprayed samples exhibited slightly lesser sliding wear resistance as compared to HVAF-sprayed samples. Two mechanisms cited for the same were (a) poorer intralamellar properties for the HVOF applied systems (P1, P2; see Table 6.6) particularly due to embrittlement occurring from carbon deficiency leading to crack nucleation and propagation and (b) near-surface plastic deformation characterized by rough wavy features formed due to shear stresses induced in the coating owing to friction between the mating parts. Interlamellar cohesion was not found to have a significant impact on sliding wear resistance. When subjected to high temperatures of around 400 °C, the wear rates increased by nearly one order of magnitude and coating systems (consisting of both fine and coarse grades) applied by diamond jet 2700 were found to have failed. The authors reasoned that in case of coatings applied using the diamond jet 2700 equipment, macroscopic cracks that were found to have nucleated and propagated down to the substrate could have been formed long before the actual test itself when temperatures would have neared the aforementioned test temperature rendering further studies on these coatings meaningless. Three possible reasons cited for the nucleation of such cracks were (a) the possibility of thermal coefficient mismatch, (b) presence of residual stresses in the coating resulting from the spray process and (c) precipitation of secondary carbides. However, the authors also noticed that only one set of the HVOF-sprayed coatings (P2W1 and P2W2) had failed. This was attributed to the huge residual stresses present in the

Table 6.5: Microstructural characteristics of the coatings produced by Bolelli et al.

Ref.	Coating system	Method	Details	Process code	Coating system	Measured parameter			
						Porosity, Vol.%	Carbides, Vol.%	$HV_{0.1}$	K_{IC}
			Powder size: Fine (W1)						
[36]	WC-10Co4Cr	HVOF	Liquid fueled (paraffin) JP 5000	P1	P1W1	3.9 ± 1.9	49.5 ± 4.7	1,400	5.45
			Gas fueled (propane) diamond jet 2700	P2	P2W1	2.0 ± 0.9	63.3 ± 9.5	1,600	6.25
		HVAF	M2 (methane)	P3	P3W1	1.1 ± 1.0	72.9 ± 2.9	1,000	6.25
			M3 (propane)	P4	P4W1	0.7 ± 0.5	54.9 ± 9.0	1,250	7.20
[37]	CrC–NiCr	HVOF	Liquid fueled (paraffin) JP 5000	P5	P5W1	3.9 ± 1.3	53.0 ± 2.7	950	3.75
			Liquid fueled (paraffin) K2	P6	P6W1	Poor sprayability			
			Gas fueled (propane) diamond jet 2700	P7	P7W1	3.5 ± 2.2	48.5 ± 3.0	920	4.13
		HVAF	M2 (methane)	P8	P8W1	6.2 ± 1.7	54.9 ± 1.7	640	2.26
			M3 (propane)	P9	P9W1	4.1 ± 1.9	50.3 ± 1.0	1,020	3.62
			Powder size: Coarse (W2)						
[36]	WC-10Co4Cr	HVOF	Liquid fueled (paraffin) JP 5000	P1	P1W2	4.1 ± 1.8	47.3 ± 5.9	1,250	4.40
			Gas fueled (propane) diamond jet 2700	P2	P2W2	7.3 ± 3.3	57.8 ± 8.5	1,100	5.40
		HVAF	M2 (methane)	P3	P3W2	1.8 ± 1.6	67.4 ± 3.3	1,000	5.00
			M3 (propane)	P4	P4W2	2.1 ± 1.7	51.3 ± 7.0	1,250	5.45
[37]	CrC–NiCr	HVOF	Liquid fueled (paraffin) JP 5000	P5	P5W2	6.3 ± 3.2	57.1 ± 0.9	840	3.59
			Liquid fueled (paraffin) K2	P6	P6W2	4.0 ± 1.7	52.9 ± 1.4	1,000	4.13
			Gas fueled (propane) diamond jet 2700	P7	P7W2	5.3 ± 2.1	49.3 ± 0.9	900	4.06
		HVAF	M2 (methane)	P8	P8W2	Poor sprayability			
			M3 (propane)	P9	P9W2	3.9 ± 1.2	53.4 ± 1.7	980	4.60

Table 6.6: Summary of tribological tests.

Ref.	Coating system	Method	Details	Process code[b]	Test method	ASTM G99[a]				ASTM G65
					Measured parameter	Wear	Friction	Wear	Friction	Volume/ mass loss
					Test temperature	RT[b]	RT[b]	HT[b]	HT[b]	
Powder size: Fine (W1)										
[36]	WC-10Co4Cr	HVOF	Liquid fueled (paraffin) JP 5000	P1	P1W1	$<10^{-7}$	0.5	$<10^{-6}$	1.1	3–4 mm^3
			Gas fueled (propane) diamond jet 2700	P2	P2W1	$<10^{-7}$	0.5	Failed	Failed	3–4 mm^3
		HVAF	M2 (methane)	P3	P3W1	$<10^{-7}$	0.5	$<10^{-6}$	0.9	3–4 mm^3
			M3 (propane)	P4	P4W1	$<10^{-7}$	0.5	$<10^{-6}$	1.1	3–4 mm^3
[37]	CrC–NiCr	HVOF	Liquid fueled (paraffin) JP 5000	P5	P5W1	$<10^{-5}$	0.7–0.75	$<10^{-4}$	0.55	60 mg
			Liquid fueled (paraffin) K2	P6	P6W1			Poor sprayability		
			Gas fueled (propane) diamond jet 2700	P7	P7W1	$<10^{-5}$	0.7–0.75	$<10^{-4}$	0.55	60 mg
		HVAF	M2 (methane)	P8	P8W1	$<10^{-5}$	0.7–0.75	$<10^{-4}$	0.55	260 mg
			M3 (propane)	P9	P9W1	$<10^{-5}$	0.7–0.75	$<10^{-4}$	0.55	60 mg

				Powder size: Coarse (W2)						
[36]	WC-10Co4Cr	HVOF	Liquid fueled (paraffin) JP 5000	P1	P1W2	$<10^{-7}$	0.5	$<10^{-6}$	1.15	5–6 mm^3
			Gas fueled (propane) diamond jet 2700	P2	P2W2	$<10^{-7}$	0.6	Failed	Failed	5–6 mm^3
		HVAF	M2 (methane)	P3	P3W2	$<10^{-7}$	0.6	$<10^{-6}$	0.85	5–6 mm^3
			M3 (propane)	P4	P4W2	$<10^{-7}$	0.6	$<10^{-6}$	1.1	5–6 mm^3
[37]	CrC–NiCr	HVOF	Liquid fueled (paraffin) JP 5000	P5	P5W2	$<10^{-5}$	0.7–0.75	$<10^{-4}$	0.55	100 mg
			Liquid fueled (paraffin) K2	P6	P6W2	$<10^{-5}$	0.7–0.75	$<10^{-4}$	0.55	70 mg
			Gas fueled (propane) diamond jet 2700	P7	P7W2	$<10^{-5}$	0.7–0.75	$<10^{-4}$	0.55	70 mg
		HVAF	M2 (methane)	P8	P8W2			Poor sprayability		
			M3 (propane)	P9	P9W2	$<10^{-5}$	0.7–0.75	$<10^{-4}$	0.55	70 mg

[a] All wear readings as per ASTM G99 are provided in mm^3/Nm.

[b] RT, room temperature; HT, high temperature (400 °C). The process codes have been changed (from the original paper) for effective comparison of the two works.

coatings applied by diamond jet spray owing to low-particle impingement velocity. For all other samples, no failure was noted albeit higher wear rates and presence of grooves coupled with abrasive scars resulting from the removal of tungsten oxides marked the wear mode at high temperatures. A fair degree of material pullout – a characteristic of adhesive wear – combined with welding of asperities was also noted resulting in high friction coefficients at these temperatures.

For abrasion wear studies, tests were carried out using a modified version of ASTM G65 using blocky shaped dry quartz sand ranging from 0.1 to 0.6 mm in size as the abrasive media at a flow rate of 25 g/min. This test involved applying a normal load of 23 N by pressing the samples against a rubber wheel possessing a surface speed of 1.64 m/s over a total sliding distance of 5,904 m. Abrasion resistance was computed using weight loss technique. In dry sand rubber wheel tests performed as per ASTM G65, a greater volume loss was noted for those coatings consisting of coarse-grained feedstock (W2 series). An interesting finding from this study was that while intralamellar factors such as carbide retention was the governing parameter for sliding wear, interlamellar attributes like cohesion and porosity were found to be the influencing factors in case of abrasive wear. These findings could be correlated to earlier jet erosion studies involving large-sized silica abrasives, where intersplat adhesion was cited to be an important factor [42].

These studies on the WC-10Co4Cr system formed the basis for further testing on the CrC–25NiCr system [37]. Once again, two feedstock powder sizes were utilized (fine ~38 μm, coarse ~45 μm). Both powder types were sprayed onto a low carbon steel substrate using five different techniques (three different types of HVOF and two types of HVAF processes). An important point noted by the authors was that out of the ten process–feedstock combinations available, only eight resulted in a satisfactory coating owing to nozzle clogging issues with the other two. One of the first important points made by the authors were on the thermal expansion coefficient aspect. It was pointed out that the thermal expansion coefficient of CrC–NiCr coatings was higher than the WC-based systems as determined from earlier studies by various other authors. The resulting coating thickness was noted to be between 300 and 400 μm.

Sliding wear tests were carried out using a ball-on-disk setup in accordance with ASTM G99 using the same parameters as in the earlier study under both room and elevated temperatures. Sliding wear tests in accordance with ASTM G99 produced wear rates in the range of 10^{-5} mm^3/N-m, which was about two orders higher compared to WC-based coating system. Furthermore, similar values were noted for coatings produced using both fine and coarse powder feedstocks. Some interesting facts pointed out by the authors were that while abrasive grooves like that produced by the ASTM G65 tests were observed, these were much narrower and shallower, indicating that the hard asperities on the alumina ball was responsible for the wear at a much smaller length scale. This was consistent with the earlier study on WC-based system by the same authors who pointed out that under sliding tests, intralamellar properties were dominant. Wear debris generated from the sliding process was also found to contribute to the grooving effect. The effect of hardness was therefore proposed by the authors to be a highly influencing factor in this type of tribo-test environment. Compared to the WC–CoCr system in [36], a transfer film was observed on the alumina counterpart in the case of CrC–NiCr and this adhesive wear could be correlated to the high friction coefficients of 0.7–0.75. Friction heating also produced a discontinuous oxide film which broke down during the sliding process and contributed to accelerating the wear. When the test temperature was elevated to 400 °C, the wear rates were found to increase by approximately half an order of magnitude. However, and when compared to the failure mode of WC–CoCr coatings, no cracks were observed in the case of CrC–NiCr coating system. The authors correlated this to the better matched thermal expansion coefficient values of the CrC–NiCr system to steel when compared to WC–CoCr system to steel. It was also pointed out that a closer thermal coefficient match would mean a lower degree of thermal stress in the produced coating rendering it more resistant to microcracking. Friction coefficient values were also found to be near the 0.55 mark, significantly lower than at room temperature, possibly due to the formation of graphite clusters, indicating the CrC–NiCr coating system to be an excellent choice for ball valves and others operating at high temperatures.

The parameters for the ASTM G65 tests were also retained as in the earlier study. As with the earlier study, dry sand abrasive wheel tests in accordance with ASTM G65 revealed abrasive grooving followed by pullout of the splats and/or Cr_3C_2 grains. As was the case with WC-based systems, interlamellar properties such as cohesion and porosity were identified to be the influencing factors. Additionally, the finer particle size feedstock was found to produce better results when compared to the coating system derived from coarse-grained feedstock owing to the higher degree of intralamellar cohesion. To conclude, it is worth mentioning that a significant difference in wear resistance of coatings produced by different processes was not observed. These findings are in contrast to the observations of Matikainen et al. [43], who have reported a significantly lower volume loss of coatings produced by the HVAF technique as compared to the HVOF method.

Similar studies involving CrC–20(NiCr) coatings were also carried out by Lih et al. [34]. The pin-on-disk tests were set up in accordance with ASTM G99 using an alumina abrasive disc as the counterface. The sliding velocity was set to 500 rpm with a wear track radius of 30 mm. A normal load of 7.84 N was applied to the pin. For the dry-sand rubber wheel tests performed in accordance with ASTM G65, silica sand (212–300 μm) was fed into the interface between the specimen and the wheel at a rate of 225 g/min. A normal load of 49 N was applied to the rubber wheel of 225 mm diameter rotating at 200 rpm. The main aim of this study was to identify the effect of particle speed and temperature on wear properties. No relation between particle temperature and wear properties were observed; however, a higher particle speed was found to produce coatings with improved wear resistance on the ASTM G65 tests and a weak dependence between particle speed and wear resistance was observed on the ASTM G99 tests. Interestingly, particle speed was not found to influence the porosity of the produced coatings, contrary to expectations. The authors went on to explain that this could have occurred due to a statistically insignificant number of samples (10 nos.) being tested in their study. However, as explained in previous works [36, 37], the better abrasion resistance for coatings sprayed with particles possessing a higher velocity could have been due to the fact that a higher kinetic energy would render a greater peening effect resulting in lower residual stresses combined perhaps with a greater degree of interlamellar bond strength. Such a relationship, however, was not evaluated in this work.

Picas et al. [6] investigated the effects of initial powder size on the tribological properties of HVOF sprayed (C-CJS [carbide jet spray] gun) CrC–25(NiCr) coatings. The microstructural characteristics of the produced coatings are summarized in Tables 6.2 and 6.3. The tests were performed under both dry and lubricated (Repsol 15W40 oil) conditions in accordance with ASTM G99 using a WC–6Co pin as the counterface. A constant load of 40 N was applied to all the coated samples. The starting powders were produced in three feedstock sizes: standard (~30 μm), fine (~10 μm) and fine (~5 μm). One of the first observations was that while the fine powders produced a much smoother coating surface finish as compared to the fine powdered feedstock, the microhardness of the fine sized feedstock coatings was found to be lower and was attributed to greater degree of decarburization. The specific wear rate of the fine particle coating was much lower than that of the standard powder coating. This was different from findings reported by Bolelli et al. [37] where no change in wear rates were observed. A closer analysis reveals that the definition of 'fine' is of importance here. The fine particles used in the earlier referenced work [37] were almost equivalent to the standard sized particles used in this study. It could therefore be argued that for a significant improvement in wear resistance, it is important that the starting powder feedstock be brought down to below 10 μm; however, further studies in this regard are required. An important conclusion drawn by the authors was that wear resistance need not be always related to the hardness with hardness itself, which is cited to be a property depending partly on intralamellar and partly on interlamellar contribution. Again, a closer look at the situation and comparison with earlier studies [37] reveals that hardness becomes more important when wear performance at higher length scales is evaluated (and would be valid only for hardness measurements using $HV_{0.3}$ or higher loading methods); for instance, tests set up in accordance with ASTM G65/ASTM G76 [33, 41]. This once again reinforces the opinion of the authors of this review work that any screening test selected for coating pre-qualification purposes must closely mirror or be indicative of the actual scenario on field; else, insignificant parameters will be evaluated and optimized leading to below par performance. The authors proposed that repeated sliding of the pin would have resulted in gradual wear out of the soft binder matrix leading to carbide pullout which could then act as third-body particles leading to an increased (standard) or decreased (fine) amount of wear depending on the particle morphology and related characteristics. At the load and sliding velocity (0.1 m/s) employed in this work, mild wear leading to a fair degree of breakout and entrapment of particles in the interface was noted as the predominant wear mechanism. This was found to somewhat agree with the wear maps described by Roy et al. [44]; a summary of which has been provided in Table 6.4. No change in friction coefficient values was noted and these values are in agreement with previous works [37].

Sidhu et al. [12] compared the tribological behavior of Cr_3C_2–NiCr and WC–Co coatings deposited by liquid petroleum gas fueled HVOF process. Both coatings were applied at a thickness of ~330 μm onto #45 alumina grit blasted T-22 boiler steels cut to pin shape with a diameter of 6 mm and length of 30 mm. The microstructure and associated properties of the produced coatings are presented in Tables 6.2 and 6.3. Pin-on-disk tests were carried out under dry conditions in accordance with ASTM G99 using a hardened steel counterpart under a normal load of 49 N at a sliding velocity of 1 m/s over a sliding distance of 4,500 m. The friction and wear graphs derived directly from the works are shown in Figure 6.9(a and b) for comparative

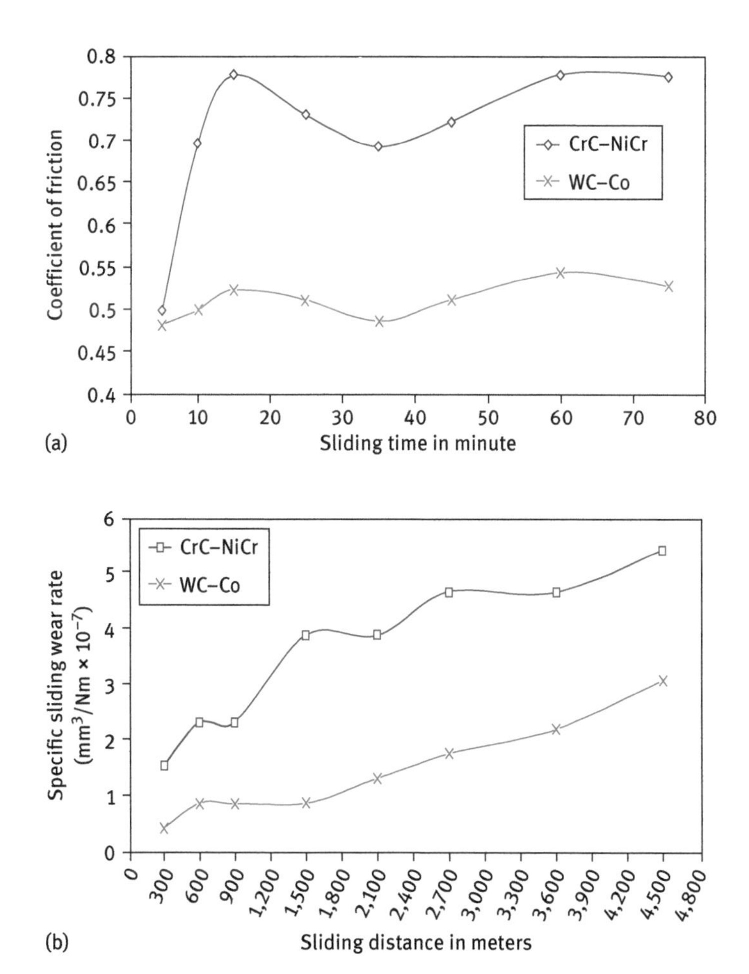

Figure 6.9: (a) Friction coefficient versus sliding time and (b) specific sliding wear rate versus sliding distance [12].

purposes. It was concluded that the HVOF-sprayed WC–Co coating exhibited much better performance, owing to its higher hardness, higher degree of uniformity and dense microstructure.

Dry abrasion performance tests have also been performed by several authors using Taber Abrasers [40, 45–47]. As in dry sand rubber/steel wheel tests, these studies can also be categorized under wear occurring at a larger length scale with the major difference being that Taber Abraser tests involve an abrasive wheel as the counterpart, whereas abrasive particles are fed into the interface of the specimen and another counterpart in case of the former.

Wielage et al. [40] compared the performance of HVOF (JP 5000; kerosene fuel)-sprayed WC–12Co, WC–10Co4Cr and CrC–25(NiCr) coatings under standardized conditions. The Taber Abrasion tests were carried out in accordance with ASTM F1978 and employed three different test wheels (H-22, H-10 and S-35) subjected to a load of 1,000 g for a total of 50,000 cycles. In cases where the wheel would not last the whole test, it was decided to use the minimum radius of the wheel as the point of stoppage of tests. However, not all combinations of test wheels and coatings were tried. In addition to drawing general comparisons on the performance of the said coatings, the effect of WC particle size (0.8 μm – fine, 2 μm – standard, 5 μm) on the wear resistance of WC–10Co4Cr coatings was also evaluated. The effect of abrasive wheels on wear performance was evaluated only for the WC–12Co system. To begin with, and as expected, all abrasive wheels abraded the WC–12Co coating differently with H-22 wheels causing the greatest degree of wear on the coating. In all cases, a sudden spike in wear rate was noted during the run-in period with wear rate values tending to a constant upon prolonged testing (read: >4 h). The tests were repeated after polishing the coatings. It was noted that the polished coatings exhibited a much lower wear rate as compared to as-sprayed ones; however, beyond 15,000 cycles, almost similar wear rates were observed. For subsequent tests, only H-22 wheels were used since this was found to produce the highest rates of wear. The effect of WC particle size on wear performance of polished coating surfaces also yielded some interesting results with the run-in period wear loss of coatings with different WC particle sizes corresponding to the weight gained via water absorption (resulting from ~65% relative humidity of the test environment). Fine particle sized coatings were found to undergo the highest amount of wear followed by standard and coarse-grained particle sized coatings. These trends were also reported elsewhere [46]. However, in both the referenced works [40, 46], the authors went on to point out that the influence of the matrix composition was a more dominant factor influencing wear resistance of coatings as compared to the carbide particle size. In a related study, Wank et al. [45] also went on to mention that the use of fine carbide particle sizes need not necessarily correspond to an enhanced level of wear resistance in all cases. In comparative tests, WC–10Co4Cr demonstrated the best performance followed by CrC–25(NiCr) and WC–12Co coatings.

Nurbaş et al. [47] compared the performance of several coatings applied by the air PS technique. A comparison has been drawn in this chapter between the results obtained for WC–12Co and CrC–25(NiCr) coating systems. All coatings in the referenced study were applied onto #36 grit alumina sand-blasted, A286 superalloy substrates at a thickness of 100–150 μm. Abrasive wear tests were carried out using a Taber Abraser fitted with a CS-17 wheel and subjected to a 1 kg load. Test data were collected and analyzed over every 1,000 cycles. One of the first observations by the authors was the higher surface roughness of the produced coatings. As compared with surface roughness values obtained by other researchers using the HVOF spray technique for depositing CrC–NiCr coatings [6, 34], those produced by the APS technique revealed higher values for similar starting powder sizes. The coatings were also found to exhibit higher porosity levels and inclusions. In general, CrC–NiCr coatings exhibited better abrasion resistance as compared to the WC–12Co coating system.

6.3.1.2 Jet erosion, slurry erosion and cavitation wear performance studies

Erosion is a common wear mode occurring in components subject to impact from entrained solids in the flowing media (which may be liquid or gas) and as such is of interest to the oil and gas (turbine and rotor blades), aerospace (space crafts) and power industries [48, 49].

Extensive studies on the mechanism of erosion occurring in as-sprayed CrC–25NiCr systems from a microstructural point of view were carried out by Matthews et al. [14]. The authors pointed out that such a study was necessary due to the high degree of contradictions found in the literature with regard to the actual mechanisms influencing erosive wear. The authors observed that earlier works did suggest three principal mechanisms to account for the influence of splat microstructure on erosion response, namely (a) microchipping and ploughing, (b) splat fracture and (c) disbondment at the splat boundaries. In other words, coatings with lower intersplat adhesive strength were hypothesized to fracture/crack during erosion. Furthermore, the combination of spray technique and starting powder size was also observed by the authors to affect erosion response. In the referenced study, coatings were sprayed at a thickness of 400 μm onto grit-blasted mild steel substrates using both HVOF and HVAF techniques. The differences in microstructure of the coatings generated through each method have been presented in Tables 6.2 and 6.3. Erosion tests were carried out in accordance with ASTM G76 using a slightly modified test setup (higher erodent velocity, smaller erodent size and larger nozzle diameter) to simulate turbine environments. The erodent used for the tests was crushed alumina with a nominal particle size of 20–25 μm with impact angle set to 90°. The erodent velocity generated was computed to be 150 m/s. Tests were carried out under both single impact (1 s) and steady-state erosion conditions of which only the former case has been detailed here. For HVAF coatings loaded under single impact, three zones from the point of impact were identified as shown in Figure 6.10(a): (a) region 1/zone 1 characterized by carbide deformation resulting from cutting by the erodent accompanied by plastic

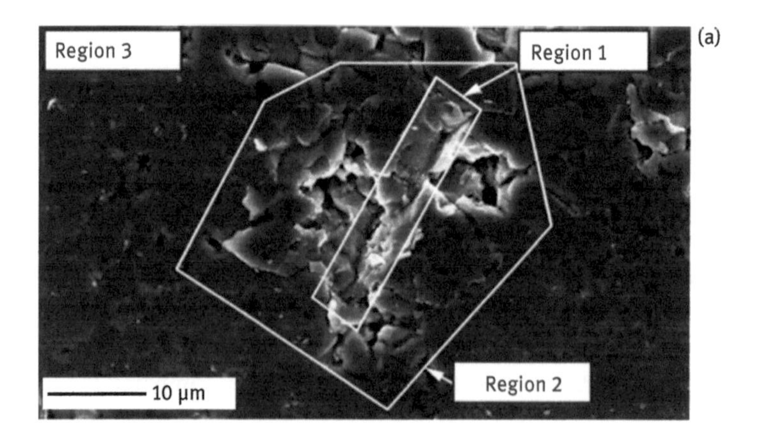

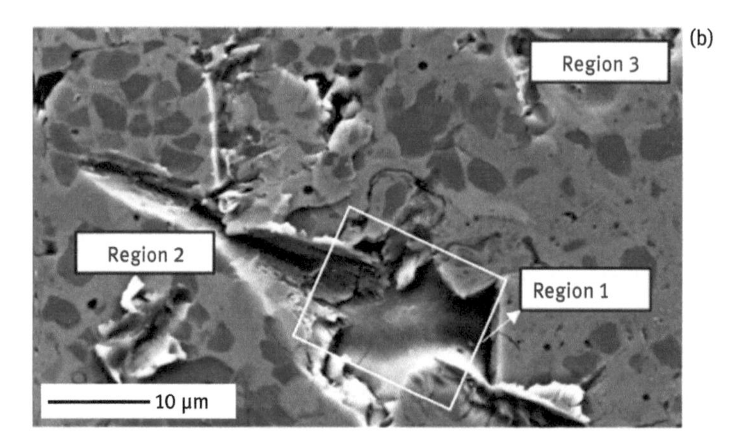

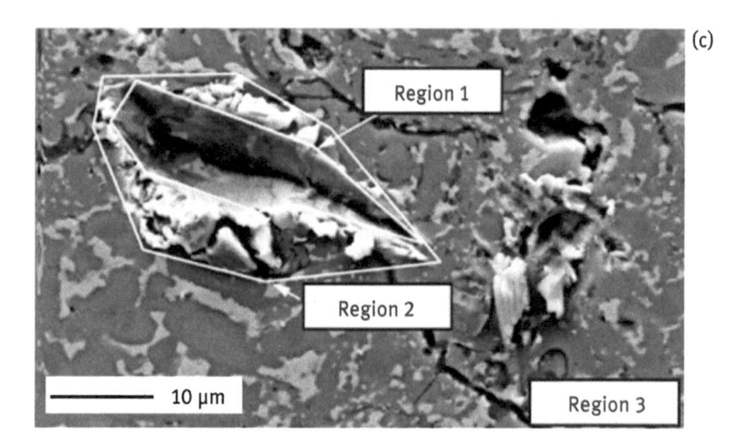

Figure 6.10: (a) Single impact erosion of HVAF coating (as-sprayed); (b) single impact erosion of HVOF coating (as-sprayed); and (c) single impact erosion of HVOF coating (heat-treated).

deformation of the matrix within the indent which was thought to have occurred due to frictional heating arising from the poor thermal conductivity of Cr_3C_2 and NiCr; (b) region 2/zone 2 was marked by extensive brittle cracking along the carbide–matrix interface with no carbide deformation as seen in zone 1. Some of the alloy binder was also found to have extruded out, owing to the relative movement of the carbide binder. It was explained that the reason for this could be due to hardening arising from some degree of carbide dissolution into the matrix along with inefficient load transfer through the contiguous carbide phases; and (c) region 3/zone 3 exhibited only a mild degree of elastic loading. HVOF coatings also exhibited similar characteristics to those sprayed using the HVAF process with region 2/zone 2 undergoing a greater amount of brittle cracking. The same is shown in Figure 6.10(b). An important point mentioned by the authors was that the amount of material loss in both cases was a direct consequence of the intersplat bonding characteristics. This was further emphasized to be a determining factor as the penetrating erodent would act as a wedge, dislodging entire splats. The intersplat strengths were also found to be adversely affected by the presence of oxide stringers in HVOF coatings (Figure 6.5; Table 6.3) acting as preferential sites for propagation of cracks. Such stringers were noted to be minimal in HVAF coatings. Compared to earlier theories in literature, factors such as the spray technique or powder feedstock size were noted to be not predominant in determining erosion resistance. As an additional supporting statement to this finding, it was also pointed out that the observations made in this work proved to be true in case of coatings applied using high-velocity processes that led to high intersplat adhesion strengths. In phase two of these works [31], the same coatings were applied using the same processes as in phase one. However, this time around, both the coatings were heat treated at a temperature of 900 °C. Once again, three zones were marked extending radially outward from the point of impact as shown in Figure 6.10(c) for clarity. The first point noted was that zones 1 and 2 were much reduced in size when compared to the earlier case. An interesting finding was that zone 2 exhibited extremely low brittle fracture. An increased amount of ductility coupled with toughening of the carbide–matrix interface was postulated to be the reason behind this. Heat treatment was also reported to have improved the intersplat bond strength. Even oxide stringers, which marked the microstructure of as-sprayed HVOF and HVAF (to some extent) coatings, were found to be lacking in the heat-treated specimens.

Jet erosion tests in accordance with ASTM G76 were also carried out by Mruthunjaya et al. [38, 39] through a series of works on WC–12Co and CrC–NiCr coatings applied on alumina grit-blasted stainless steel substrates. In the first of these works, WC–12Co coatings produced from feedstock of three different particle sizes (25, 39 and 68 μm) were applied using the HVOF technique. An air jet erosion tester was used to force a stream of abrasive particles at a velocity of 60 m/s for 10 min at different impact angles, namely, 30°, 45°, 60°, 75° and 90°. The experiments were conducted at 800 °C. In all cases, the erosion rate was found to follow a bell-shaped curve with the maximum erosion rate of ~425 mg/kg occurring for the fine mesh powder coating (25 µm) at a 60° impingement angle. In phase two, tests on CrC–NiCr coatings were carried out using the same setup albeit at a lower temperature of 600 °C for impingement angles of 60°, 75° and 90°. An important point to note is that three different compositions of the hard metal coating with the same average particle size of 25 µm were used, namely, CrC–20NiCr, CrC–12NiCr and CrC–8NiCr. CrC–20NiCr was found to produce the best performance with a maximum erosion rate of ~275 mg/kg occurring at 75° impingement angle.

Erosion can also result from the flow of entrained solids in a liquid media. As explained earlier, in a typical industrial setting, a combination of wear modes often comes into play and as such, a coating should be designed to withstand a multi-wear mode scenario. Such a phenomenon involving both erosion and cavitation is more commonly experienced by fluid handling equipment, and works in this regard have also been carried out by several researchers [41, 50, 51]. Cavitation in pumps and turbines resulting from cyclic stresses imposed by the microjetting action of imploding droplets owing to a reduction in pressure along the fluid flow path has been a major area of study in recent years with thermally sprayed coatings being widely employed to protect and extend the working life of equipment [52–54]. Santa et al. [50] reported that the predominant wear mechanism taking place in thermally sprayed coatings under cavitation conditions (tests were set up in accordance with ASTM G32) was the brittle fracture of hard phases, and fatigue failure of the ductile binder phase while microploughing and cutting of the softer phase was noted to be the wear mechanism governing slurry erosion performance. Sugiyama et al. [51] pointed out that the wear resistance of coatings under slurry erosion conditions (calculated using the reciprocal of the volume loss rate or time taken to wear out 1 mm^3 of volume of the coating) was influenced by the hardness. An important relation pointed out by the authors in this work was that the measured Vickers Hardness readings ($HV_{0.2}$) were found to increase by the third power to wear resistance of coatings under slurry erosion tests carried out at 60° and 90° impingement angles, whereas at a 15° impingement angle, the same variables were related by the fourth power. Lima et al. [41] carried out extensive tests on WC-based coating systems under cavitation wear mode. Fracture toughness was directly proportional to the cavitation resistance of the tested thermally sprayed coatings. An important point mentioned by the authors was that owing to the anisotropic properties of the thermally sprayed coatings, it was necessary that fracture toughness measured via indentation was recorded parallel to the coating–substrate lay. The adopted thermal spray method, addition of interlayers and so on were found to be less influencing factors controlling cavitation resistance. The size of the wear debris was also found to be related to the fracture toughness with tougher coatings producing smaller wear debris and in turn rendering the coating more resistant to cavitation erosion. Other authors have also mentioned that coatings with a lower pore density produce higher cavitation resistance [51].

6.3.2 Coating performance enhancement techniques

In the earlier section, majority of the works that were reviewed involved tests on coatings in "as-sprayed" condition. However, in such a state, the coating microstructure is prone to have defects such as high porosity, low toughness and intralamellar cracks. Over the years, researchers have invested a great deal of time and effort in their pursuit to overcoming such limitations associated with CrC–NiCr coatings. Simultaneously, researchers have also focused their efforts on imparting special properties to the coating such as self-lubrication, with a view to expanding the range of applications possible with CrC–NiCr cermet coatings. This section is intended to highlight recent developments in the field of CrC–NiCr coatings' performance enhancement and testing along with identifying directions for future research. Due to the widely varying nature of works, this section has been subdivided into two parts to ensure clarity. In part 1, activities carried out prior to the coating application such as grain size selection, addition of fillers into the powder feed, process changes such as addition of transition layers (interlayer) have been discussed. In part 2, post-coating activities such as laser posttreatment have been discussed.

6.3.2.1 Effect of pre-coating parameters/activities on coating performance

Effect of grain size

The influence of grain size on tribological properties was extensively studied by Roy et al. [44]. As stated by Roy et al. [55] in a related study, the main differentiating factor between nanostructured and conventional coatings is the fact that in the nanostructured coatings, the biggest crystal could have the size of a grain up to ~100 nm. In the work referenced earlier [44], both conventional (grain size > 1 μm) and nanocrystalline (<100 nm) coatings were applied using an HVOF system onto #13 alumina grit-blasted mild steel substrates. Image analysis revealed 1.5% and 2% porosities combined with 0.5% and 0.75% oxides in the nanograined and conventional coatings, respectively. A three pin-on-disk test rig, with 100Cr6 steel counterpart, was used to carry out the tribo-tests under varying p–v conditions. The findings of the authors are summarized in Table 6.7.

In general, at low loads, it was observed that nanograined coatings demonstrated a much lower coefficient of friction as opposed to conventional grained coatings. At high loads of 140 N, however, values indicating the possibility of seizure were observed for both types of coatings. Three reasons were put forward by the authors for the lower friction coefficient demonstrated by the nanograined coating system: (a) the lower surface roughness by almost 40% as opposed to conventional grained systems, (b) the increased hardness of the nanograined system and (c) greater grain boundary energy of the nanograined system indicating a higher degree of amorphousness. Further analyses of the data at 70 N load for nanograined coating show that the lowest friction occurred for intermediate sliding velocities. Examination of

Table 6.7: Summary of test results.

Grain size	Conventional grains			Nanocrystalline grains		
Property	**Friction coefficient**			**Friction coefficient**		
Load/sliding speed	35 N	70 N	140 N	35 N	70 N	140 N
0.125 m/s	Not presented	0.70	Not presented	Not presented	0.70	Not presented
0.25 m/s	0.2	0.53	>1	0.1	0.40	>1
0.5 m/s	Not presented	0.70	Not presented	Not presented	0.55	Not presented
Property	**Wear**			**Wear**		
Load/sliding speed	35 N	70 N	140 N	35 N	70 N	140 N
0.125 m/s	Mild	Breakout	Breakout	Mild	Delamination	Delamination
0.25 m/s	Mild	Breakout	Seizure	Mild	Delamination	Seizure
0.5 m/s	Breakout	Breakout	Seizure	Delamination	Delamination	Seizure

worn surfaces showed that there was a higher rate of wear of the binder matrix in case of conventional grained coatings as compared to nanograined coatings at a load of 70 N. However, and at a load of 140 N, severe plastic deformation accompanied by adhesive pullout of the materials was observed in line with the high friction coefficients. The reason for high friction coefficient at high loads was attributed to the rise in contact temperatures resulting in softening of the material accompanied by an increase in the real contact area. A higher amount of fluctuation in the graph was also noted pointing to the occurrence of stick-slip phenomenon and/or wear debris particles. An important finding from the work was that in case of loading at 70 N (0.25 m/s sliding velocity), cracks in the nanograined coating nucleated at the subsurface at a characteristic depth; however, for the conventional grained coating, the cracks were formed due to breakage of carbide particles further to wearing out of the soft matrix. Cross-sectional images of the same derived from [44] are presented in Figure 6.11(a and b).

Another interesting point explained by the authors was that no phase transformations were noted in either coating in their work though an earlier work did cite the formation of the $Cr_{23}C_6$ phase. It was explained that this could have occurred due to the higher residence time of the particles during the spray process in this work as opposed to the earlier work. A qualitative assessment of wear was carried out. The authors concluded that wear mechanisms in nanograined coatings were either mild wear, delamination wear or seizure wear, whereas the mechanisms in conventional grained coatings were mild wear, breakout wear or seizure wear for low, medium and high p–v regimes.

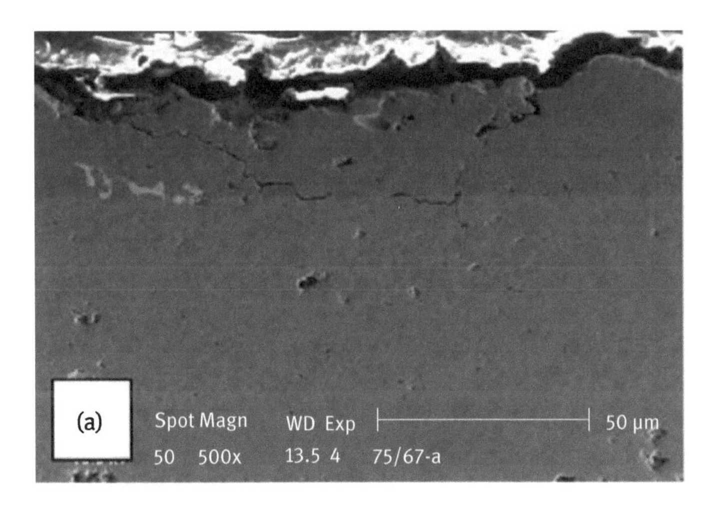

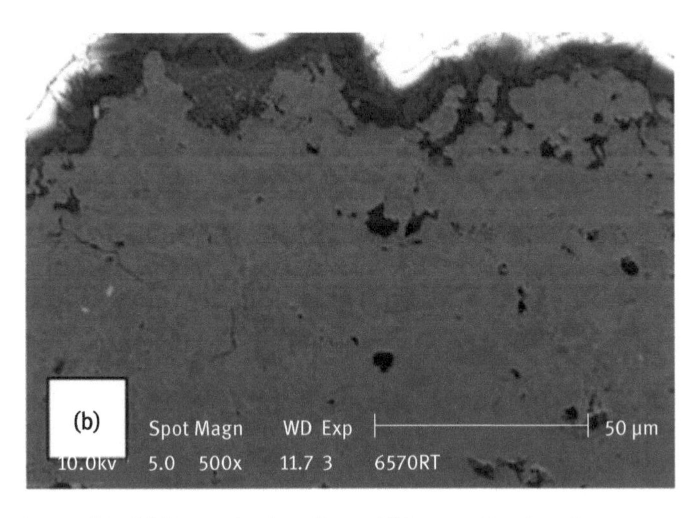

Figure 6.11: (a) Nanograined coating and (b) conventional coating.

Kai et al. [25] created a CrC–NiCr coating from a Ni–Cr–C powder as atomized (Ni: 52.26, Cr: 45.88, C: 1.86) and as cryomilled using HVAF process. Fracture toughness studies were carried out using the nanoindentation technique. While the conventional coating (possessing large grained microstructure) demonstrated a slight crack at a load of 10 N along the grain boundary, the nanostructured coating showed no such effects. The nanostructured coating was reported to possess a better fracture toughness than the conventional coating while also possessing more hardness and reduced porosity levels.

In a separate study by Picas et al. [33], the mechanical and tribological properties of conventional and nanocrystalline CrC–NiCr coatings in two compositions and three grain sizes – CrC–25(NiCr) standard, CrC–25(NiCr) nanostructured and CrC–40(NiCr) nanostructured – applied using HVOF at a thickness of ~30 μm on steel substrates were compared. The higher carbide content of the CrC–25(NiCr) coatings rendered it harder than the CrC–40(NiCr) coating. However, between the standard and nanostructured CrC–25(NiCr) system, the standard coating had the higher hardness and was attributed to a lower degree of decarburization and decomposition of the carbides during the spray process. Decarburization in nanograined coatings has also been reported by Racek [56]. The surface roughness of the nanostructured coatings was also noted to be a tad lower than that of the standard coating. Tribological tests were set up in accordance with ASTM G99 using a WC–6Co ball as the counterface under normal loads of 30 and 40 N over 10,000 cycles at a velocity of 0.1 m/s. For nanostructured CrC–25(NiCr) coatings, friction coefficient values of 0.38 (30 N) and 0.34 (40 N) were reported; however, CrC–40(NiCr) coatings demonstrated lower coefficient of friction values at 0.13 (30 N) and 0.25 (40 N). The previous work by Roy et al. [44] hadn't reported the friction coefficients under these conditions. In general, CrC–40(NiCr) coatings were found to possess a greater wear resistance as compared to CrC–25(NiCr) coatings and this was hypothesized to be due to the better intersplat adhesion of the CrC–40(NiCr) coating system. A comparison with the wear maps drawn out in the earlier work [44] shows that only mild wear should have occurred for the CrC–25(NiCr) nanograined coating; however, the SEM images presented in this work showed characteristic delamination of the coating. CrC–40(NiCr) exhibited greater plastic deformation and subsequently no coating delamination in better agreement with the aforementioned wear map.

Addition of fillers to create self-lubricating coatings

Graphite and carbon nanotubes (CNT) as coating fillers have been popular choices due to their lubrication properties and several researchers in recent years have used the same to improve the tribological properties of various generics of coatings [57]. Various other fillers have also been used and are described in this sub-section.

Corte and Sliney [58], and Sliney [59] have carried out extensive works on adopting silver (Ag) and CaF_2/BaF_2 as fillers to modify Cr_3C_2-based hard metal coatings due to their exceptional lubricating properties and thermal stability. The main aim of their work [58] was to determine a satisfactory tribopair for piston ring–cylinder applications. The research work was conducted at NASA Lewis Research Center. Ag was chosen due to its low shear strength and the CaF_2/BaF_2 eutectic (62:38 wt.%) owing to its ability to provide lubrication at temperatures exceeding 500 °C. Furthermore, the authors also went on to mention that the fillers were chosen based on their individual abilities to operate effectively at different temperature ranges – silver in low temperature zones (running-in period) and the eutectic at higher temperatures. The tests were carried out under hydrogen and helium atmospheres (each with over 99% purity and routed into the chamber at 0.014 m^3/min) over a wide temperature range of 25–760 °C to simulate the working environment of a Stirling engine (especially in the case of hydrogen environment). Initially, powder blends of Cr_3C_2–NiCoAl, Ag and CaF_2/BaF_2 in various compositions were prepared. These were then plasma sprayed onto a disk (made of special high-temperature alloys) and tested under a load of 0.5 kg at 2.7 m/s sliding velocity. Counterfaces were changed after each trial and the test was repeated to determine the optimum tribopair. The tests revealed that a 70 wt.% metal-bonded Cr_3C_2 blended with both Ag and CaF_2/BaF_2 in a 1:1 ratio (15 wt.% Ag to 15 wt.% CaF_2/BaF_2) produced the best results. As expected, for the coating blend stated earlier combined with a counterface of hardenable cobalt–chromium alloy (Co-59 wt.%, Cr-30 wt.%, W-4 wt.%, Ni-2 wt.% and others), the friction coefficients recorded were in the range of 0.2 ± 0.05 (at 350 °C) with extremely low wear rates. A sufficient lubricant film was observed in between the mating parts leading to a low degree of wear and friction coefficient. Furthermore, and as opposed to the unmodified coating sample, no pullout of carbide phases (leading subsequently to abrasion of the coating) was observed in the modified coating.

Bartuli et al. [60] developed a self-lubricating NiCr–CrC coating by embedding graphite in the matrix (by mechanical mixing). The feedstock used consisted of nickel-cladded graphite to avoid the decomposition of graphite during the plasma thermal spray process. The final coating showed an effective presence of graphite within the coating uniformly distributed among cermet lamellae. It was postulated that the incorporation of graphite in the matrix would lead to the elimination of a requirement for external lubricant (e.g., oils) under high load conditions accompanied with relative movement between mating surfaces which would in turn prevent the generation of hazardous industrial waste. Tribological characterization of the system (coating on steel substrate) was carried out under dry conditions using a pin-on-disk test rig to determine the optimum concentration of graphite within the matrix producing the least friction and wear coefficient values. It was concluded that a graphite concentration of >25 vol.% was necessary to produce a significant improvement in tribological performance. Furthermore, the optimum values of both friction and wear coefficients were observed at a filler loading of 35 vol.%.

In another study, Singh et al. [61] prepared CNT-reinforced Cr_3C_2 coatings using the APS technique. To increase the enthalpy of the plasma plume, hydrogen gas was employed. The CNT and Cr_3C_2 powders were initially blended at <50 rpm using two different concentrations (1 and 2 wt.%) for a period of 7 days to ensure homogeneity. No binder phases were used in this study. Elastic modulus and hardness properties evaluated using nanoindentation studies (Oliver and Pharr method) revealed notable improvements in the CNT-reinforced Cr_3C_2 matrix. A pileup of dislocations around the intender showed that the CNTs contributed to accommodating the plasticity. An increase in microhardness ($HV_{0.3}$) was also noted and was hypothesized to be due to the second phase particles (CNT) acting as anchors preventing intersplat movement during indentation. Pin-on-disk tests were also carried out under dry conditions using a 100Cr6 steel static counterpart at a normal load of 10 N at a sliding speed of 15 cm/s. Coefficient of friction values was noted to initially rise to a high value in case of the Cr_3C_2–2 wt.% CNT composite possibly due to interlocking of the asperities on the counterpart with the CNT. However, with sufficient run-in, the friction coefficient values steadied at around the 0.65 mark. This was postulated to have occurred due to the smoothening out of surface asperities with sufficient run-in and due to the lubricating effect of the CNT reinforcement. The Cr_3C_2–2 wt.% CNT composite however exhibited the best wear resistance (wear rate ~3.6 mg/km) by approximately 45%. This was attributed to a combination of the lubrication effect as also the intersplat anchoring by CNTs.

Addition of transition layers

Zang et al. [16] studied the effect of having a smooth gradual change in hardness between the substrate and the coating layer. The transition layer selected was Ni45 deposited by a laser cladding process. Ni45 was reported to be selected as the transition coat owing to its medium hardness (higher than the selected substrate [20Cr2Ni4A] and lower than the coating) and compatibility with both the substrate and the coating material. Another important reason stated by the authors was that Ni was known to be efficient in reducing thermal stresses. A fiber laser was used at 3 kW and 1,070 nm wavelength to deposit the particles on a 20Cr2Ni4A substrate sample. The final produced coating showed a smooth and gradual change in microhardness from the substrate ($HV_{0.2}$ ~400 at substrate–Ni45 interface) to the top CrC–NiCr layer ($HV_{0.2}$ ~650 at Ni45–top coat interface) resulting in avoidance of stress concentration within the coating. These properties were also found to positively impact the tribological characteristics of the system. For the tribo-tests, a ball-on-plate test rig was used with a load of 30 N being applied by a SiC ball of 10 mm in diameter over a stroke length of 1,000 µm at a frequency of 50 Hz for a period of 30 min. The friction and wear tests were repeated at temperatures of 20, 100 and 300 °C. X-ray diffraction analyses revealed no $Cr_{23}C_6$ or oxide phases in the coating. Most of the phases were Cr_7C_3 and Cr_3C_2, and the reason stated by the authors was that it occurred possibly due to the process of rapid heating and cooling during the laser cladding process. Friction tests were carried out on both bare and coated substrates. The precipitation of the Cr_7C_3 phase has been cited by some authors to improve the wear resistance on account of its higher hardness [62]. At lower temperatures, the mechanism of wear was noted to be of abrasive mode for both coated and bare substrates, whereas softening at high temperatures resulted in an adhesive wear mode of the substrate while the CrC–NiCr system was found to be still undergoing abrasive wear. Energy-dispersive X-ray spectroscopy analyses also showed the presence of oxygen hinting at the possibility that the oxide layer formed at higher temperatures would have contributed to reducing the friction coefficient.

6.3.2.2 Effect of postcoating activities on coating performance

Laser posttreatment

As discussed earlier, the use of laser has been associated with improving the performance of various cermet coating systems [9]. This process, also referred to as a remelting operation, seeks to reduce microstructural defects (such as porosity) and increase the microhardness, thereby rendering a high-performance coating [35]. The reduction in the corresponding loss volume was 35–65%. The coating microhardness was also observed to have increased due to the eventually increased contiguity of the carbide network. The laser remelting process was also reported to possess a massive advantage over the conventional heat-treatment method (using an oven), one of the major ones being shorter treatment time. The effects of heat treatment on mechanical and tribological properties of CrC–20(NiCr) coatings have also been reported by Picas et al. [63]. Optimum wear resistance properties were achieved at 900 °C and were postulated to be due to the increase in hardness caused by the precipitation of fine carbides in the matrix. As may be noted later in this chapter, the effect of hardness on wear resistance becomes more significant when the evaluation of wear resistance takes place on a larger length scale. Rauf et al. [64] proposed the application of Ni–20Cr coatings for protection of boiler tubes and reported improvements in the oxidation resistance caused by reduction of the porosity and splats inside the coating further to laser posttreatment. As with any other thermal process, laser cladding whether used for material deposition to form the coating layer or as a posttreatment (remelting) process still has several controllable parameters that may affect the final produced coating. Those parameters can be optimized using design of experiments. Laser power, inert gas pressure, defocus distance and scanning speed were reported to be influencing parameters in works carried out by Rauf et al. [65] on Ni–20Cr coatings.

Mateos et al. [35] investigated the tribological behavior of plasma sprayed, 400 µm thick CrC–25NiCr cermet coating on shot-blasted, preheated, AISI 1043 steel blocks by using a block-on-ring-type linear contact test rig in accordance with the requirements of ASTM G77 standard under both dry and lubricated conditions. The ring counterpart was manufactured from tempered and hardened AISI 1043 steel. For lubricated tests, SAE30 oil was used and added dropwise into the contact zone between the mating bodies. The sliding distance in case of dry and lubricated tests was set to 6,300 and 15,700 m, respectively. Furthermore, one set of coatings was laser treated using a CO_2 laser. Porosity of the coatings was measured in accordance with ASTM 562. Cross-sectional optical micrographs shown in Figure 6.12 reveal that the porosity levels went down from ~9% to nearly 0% further to laser posttreatment.

The laser posttreated coating system also fared better in adhesive strength tests carried out in accordance with ASTM C633. It was postulated that the reason for this could have been due to a fair degree of metallurgical bonding with the substrate. Dry sliding tests were initially carried out using a normal load of 333 N and sliding velocity of 0.4 m/s. The major finding from this study was that two distinct zones of wear were present: (a) zone 1 corresponding to a run-in period where a great degree of material loss took place via spalling and due to the action of wear debris particles; (b) zone 2 corresponding to a steady-state period where wear volume was found to increase gradually. The wear rates under each of the aforementioned zones are presented in Figure 6.13.

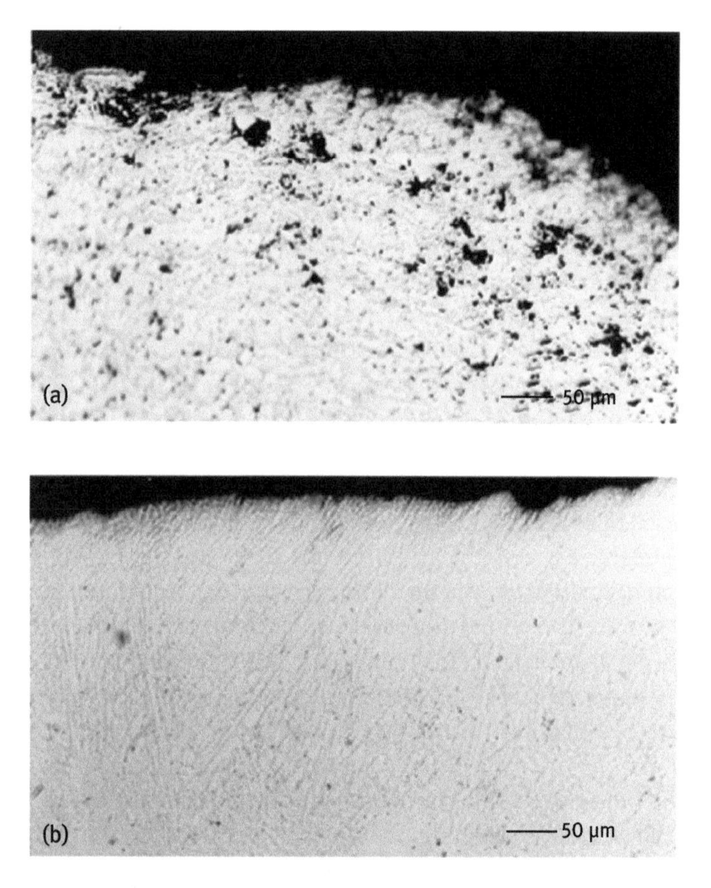

Figure 6.12: Optical micrograph of (a) as-sprayed coating and (b) laser remelted coating [35].

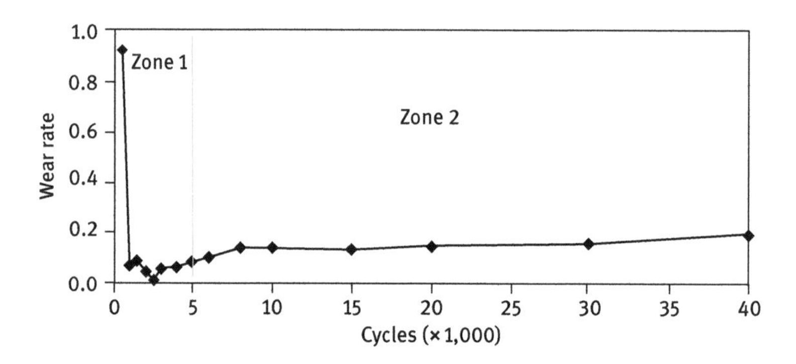

Figure 6.13: Wear rates and behavior in various zones.

The authors identified this to be in conformance with the Czichos model, which states that in the run-in period (zone 1), the *square of the wear volume is proportional to the sliding distance (cycles)* and further to that (zone 2), the *wear volume increases linearly with sliding distance (cycles)*. These were found to hold true for a range of other loads too. Specific wear rates (wear loss volume per unit of sliding distance per unit load) were measured only for the steady-state period of the tests at various loads and were found to be ~5.3×10^{-6} mm^3/Nm. The amount of wear was found to increase linearly with increasing load. Similar trends could not be picked up on the relation between wear volume and sliding speed. It was argued however that below a threshold value, the oxide film would play a role in protecting the surface; however, beyond the threshold value, high interface temperatures would result in the oxide film breaking down. For lubricated contact tests, wear volume was found to vary linearly with sliding speed for tests carried out at different loads in the order of 1,400–2,700 N. The average friction coefficient value recorded under dry conditions was ~0.49 while under lubricated conditions, the same was observed to be ~0.078.

Janka et al. [13] evaluated the effect of laser posttreatment on the tribological behavior of CrC–NiCr coating system sprayed using a paraffin-fueled HVOF gun (K2) onto a steel grit-blasted surface at a thickness of 350 µm. The microstructure of the as-sprayed coating system is presented in Tables 6.2 and 6.3. Laser posttreatment had a substantial impact on the microstructure. Posttreatment, it was noticed that there was precipitation of secondary carbides accompanied by transition from solid-solution hardening to precipitation hardening of the binder matrix. There was also a notable increase in hardness of the coating that resulted from the increased contiguity of the carbide phase. Finally, the precipitation hardening phenomenon also resulted in increased fracture toughness. Comparisons were drawn to a WC–10Co4Cr coating system sprayed with the same process. The CrC–NiCr coating was then irradiated using a high-power laser (3 kW) in continuous-wave mode using a wavelength of 975 nm at various travel speeds. It was ensured that the maximum surface temperature was kept to below the melting point of *fcc*-NiCr at 1,400 °C. Coatings remelted using fast travel laser speeds were found to have cracked. Furthermore, the microstructure of the coatings remelted using the slowest travel speed was found to have an Ostwald ripened structure. Abrasive tests were carried out using the ASTM G65 procedure under two types of loading regimes: (a) low stress loading characterized by the abrasive particles not getting affected by the loading process owing to one of the bodies absorbing the energy via elastic deformation and (b) high stress loading regime characterized by the fracture of the abrasive particles. For low stress loading regime, a rubber wheel was used to apply a 130 N load over a sliding distance of 2,100 m at a sliding velocity of 2.4 m/s. The abrasive, Ottawa silica sand (200–300 µm) flow was set to 330 g/min. For high stress loading regime, a Hardox 500 steel was used to apply a 45 N load over a sliding distance of 300 m at a sliding velocity of 1 m/s. The abrasive flow in this case was set to 150 g/min. Under both regimes, the laser-treated CrC–NiCr coatings were found to fare much better than the as-sprayed coating. The recorded performance in terms of volume loss is presented in the graphs taken directly from the referenced work (see Figure 6.14(a and b). Under high stress regimes, the as-sprayed coating microstructure revealed microcracks that were found to propagate through the binder matrix and carbide phases. The more pronounced improvement noted under high stress regime was attributed to an enhancement in fracture toughness of the binder matrix (Figure 14(a)). This was pointed out to be because the laser treatment was primarily meant to affect the properties of the binder phase only, whereas in the low-stress regime, most of the load was indicated to be borne by the hard phases only.

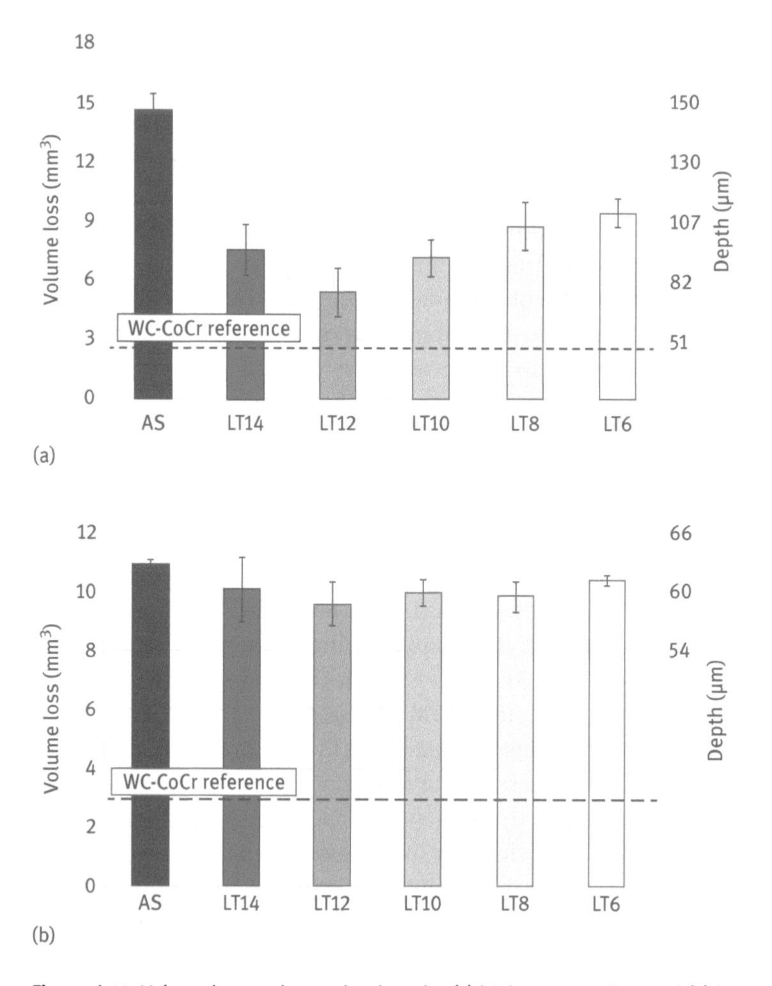

Figure 6.14: Volume loss and wear depth under (a) high stress regime and (b) low stress regime [13].

6.3.3 Coating performance under fatigue loading

Metallic components in industrial equipment are often subject to cyclic stresses, which may be mechanical, thermal or electrical in nature. Fatigue and fracture studies are therefore essential in painting a complete picture of the coating performance. Fracture response of CrC–NiCr coatings has been reported by certain authors through modeling software such as ANSYS [66]; however, more extensive studies on fatigue performance have been conducted in recent years. Zhang et al. [32, 67, 68] carried out extensive studies on the failure mechanisms and durability of plasma-sprayed CrC–NiCr coating under rolling contact conditions. As mentioned in their works, the specific failure mode, referred to as rolling contact fatigue (RCF), is generally characterized by cracking and pitting or delamination in the sublayer near the contact surface of the bodies rolling against each other. The test was conducted by depositing CrC–NiCr particles using PS on a ring-type geometry substrate of commercial medium carbon steel. In the cited work [67], the coating specimens were held in a collet and loaded against AISI 52100 balls. The RCF test was conducted under lubricated conditions using SAE 46 oil lubricant. Four different loads (50, 100, 200 and 350 N) were applied. The configuration of the test-rig was such that under the applied loads, equivalent Hertzian contact pressures of 1.507, 1.898, 2.391 and 2.882 GPa were experienced by the counterface balls. One of the first observations of the authors was the presence of intralamellar cracks and these were reported to have occurred due to the presence of residual stresses resulting from coating shrinkage. SEM scans of the as-sprayed coating indicated in Figure 6.7(b) against [32] in Table 6.3 clearly reveal the presence of the said cracks. The RCF failure modes observed were classified into four main categories: surface abrasion, spalling, cohesive delamination and interfacial delamination with spalling, cohesive delamination and interfacial delamination being predominant fatigue failure modes for CrC–NiCr cermet coatings. At contact stresses of 1.507 and 1.898 GPa, spalling and cohesive failure were noted to be predominant failure modes, whereas at higher stresses spalling was accompanied by interfacial delamination. It was pointed out that the depth of spall nucleation via microcracking was much shallower than the computed depth of maximum shear stress. An interesting finding from the referenced work [68] was the correlation between the cohesive/interfacial failure modes and the location of three predominant factors, namely (a) orthogonal shear stress, (b) maximum shear stress and (c) maximum stress amplitude with the location of maximum stress amplitude being of prime importance. It was observed that at lower stresses, the location of the maximum stress amplitude lay within the coating thereby promoting cohesive failure whereas at higher loads, the location of the maximum stress amplitude was found to have shifted into the substrate or to the coating–substrate interface leading to interfacial delamination. Cohesive delamination was also found to have occurred in certain cases involving higher stresses where the coating thickness was found to have exceeded the control values leading to the location of the maximum stress amplitude to lie within the coating. With increasing contact stresses, life parameters such as L_{10}, La and L_{90} were found to decrease (L_{10} represents the lifetime when 10% of the specimens have failed). The fatigue failure mechanism of the coating under rolling conditions was reported to be related to the coating microstructure and shear stress location. Previous studies showed that coatings deposited by different techniques resulted in varied behavior in RCF tests. Nieminen et al. [69] reported that HVOF demonstrated better performance during the RCF test as compared to those applied using the APS technique. The higher velocity of spray particles produced by HVOF was cited as the primary reason for this improved performance. Varis et al. [70] investigated the parameters and performance of the HVOF process in producing CrC–NiCr coating and its effect on fatigue performance. The coating was done on flat carbon steel (S355) substrate with CJS at different values of HVOF fuel (kerosene) and oxygen flow rates in addition to one trial set with a different HVOF system known as diamond jet hybrid. An in situ coating property sensor (a sensing device that allows the evaluation of the various stresses and deflections encountered by the substrate during the spray process) for residual stress measurements was also installed. It was found that from the point of spray initiation to the end of the spray cycle, the stress developed on the coating moved from tensile to compressive depending on the temperature differences and peening effect on the substrate. The final residual stress, however, was observed to be compressive. The evolution of high compressive stress (desirable) was achieved only by CJS which proved that the spray technology adopted has a significant impact on residual stress and hence the fatigue performance. It was thus reported that a high impact energy of coating particles is preferred to produce a relatively deep compressive residual stress layer on the substrate, which would subsequently hinder crack initiation and growth. It was also found that higher surface roughness values decreased the fatigue performance as notch-like surface structures involved could promote cyclic slip and fatigue crack nucleation. Authors of both works discussed earlier [69, 70] reported limitations with the PS technique in resisting early fatigue failure. On the other hand, both papers suggested employing the HVOF technique to improve fatigue life.

Agrianidis et al. [71] compared the fatigue resistance of WC–CoCr, CrC–NiCr and Ni20Cr coatings deposited on P91 substrates using a novel impact loading technique. It was pointed out by the authors that while conventional tribological tests, namely pin-on-disk, scratch tests, microhardness and so on, could give an indication of the tribological properties, fatigue properties would not be revealed. In their works, fatigue conditions were simulated using a carbide ball set to oscillate and impact the coated substrate at a frequency of 50 Hz. Microcracks similar to those generated in earlier works [67] were noted to occur in CrC–NiCr coating systems leading to spalling. Interfacial delamination was also found to have occurred in CrC–NiCr coatings. Spall features coupled with abrasive wear marks were also noted on WC–CoCr coatings; however, the no cohesive delamination was found to have occurred and this was attributed to the higher fracture toughness of the coating. A comparison of endurance curves of all three coatings revealed WC–CoCr to be the best performing coating system followed by CrC–NiCr and Ni20Cr.

6.4 Summary

In this work, the authors have carried out a detailed review of the studies carried out in recent years on CrC–NiCr coating systems. Advancements in thermal spray processes for application of CrC–NiCr systems have been briefly discussed. Recent developments in CrC–NiCr systems have also been explored in detail. Various mechanisms responsible for the exceptional performance or failure of CrC–NiCr coatings under a spectrum of different tribological environments have also been discussed. The tribological properties and failure mechanisms of various coatings applied using a variety of techniques have been studied extensively by several authors and have been reviewed comprehensively in this work. In each case, detailed descriptions of the test conditions and counterparts have been provided to help the reader visualize the tribo-environment and also due to the fact that a result obtained under one environment can be drastically different from another. Comparisons have been drawn, where appropriate, to similar tests carried out on WC–CoCr/WC–Co coatings. As explained by Janka et al. [13], CrC–NiCr coating would be an ideal substitute to WC-based cermet coatings owing to its lesser cost, easy availability, closely matched thermal coefficient to steel, higher resistance to chlorides, alkalis and acids combined with exceptional performance at high temperatures. Lima and Marple [72] carried out an extensive review on thermally sprayed nanostructured coatings. It was reported that coatings produced by spraying nanosized ceramic particles, namely, yttria-stabilized zirconia, titania, alumina, alumina–titania composites, reported exceptional tribological performance. While our analyses of studies conducted on advanced coatings and spray processes reveal that works involving carbon-based fillers are available, much research into the tribological performance evaluation of hybrid CrC–NiCr coatings produced by infusing novel performance-enhancing fillers such as graphene or other nanostructured ceramics remains to be carried out.

To conclude, it may have to be argued that even though a linear one-to-one mapping of all microstructural properties to tribological performance isn't possible, it would be ideal to have some of the characteristics presented in Table 6.8. The authors have tried their best to determine common findings from various scientific papers for the reader's interest; however, it is best that each is validated through further testing/ research:

Table 6.8: A linear mapping of microstructural features and other properties to desired performance.

Desired property	How do we achieve it?	Effect	References
Dense microstructure or reduced porosity	Use proper spray technique	Better interlamellar properties result in improved performance on higher length scale tests such as ASTM G65	[36, 37, 42]
	Laser posttreatments	Increased densification results in better tribological performance	[9, 35, 64]
Erosion resistance	Use proper spray technique and powder size combination	Improved fracture toughness leading to better erosion resistance	[14]
	Post heat treatment	Increased intersplat bonding leading to better erosion resistance	[31]
Fatigue performance	Use nanograined feedstock	Increased fracture toughness and improved fatigue performance	[25]
	Increase hardness (refer to hardness measurement considerations reported in [37])	Improved cavitation resistance (tested alongside fatigue performance)	[41, 51]
Hardness	Hardness cannot be directly correlated to wear performance in case of thermally sprayed cermets and several conflicting reports in literature are observed. This may be primarily due to some contribution of hardness value coming from intralamellar and other from interlamellar properties. $HV_{0.1}$ (intralamellar property) $HV_{0.3}$+ (interlamellar property) Additionally, different materials exhibit different tribological performance depending on the testing length scale	Improved performance on higher length scale tests such as ASTM G32 or ASTM G65 (while noting that hardness should be measured using the relevant method that stresses more than one lamellae such as in $HV_{0.3}$ or $HV_{0.5}$ measurements [37])	[33, 41]
	Add transition layers	Improved tribological performance	[16]
Increased peening effect	Increase the particle impact speed on the substrate	Improved performance on higher length scale tests such as ASTM G65	[36, 37]
		Improved fatigue performance	[70]
		Improved performance on higher length scale tests such as ASTM G65	[36, 37]

(continued)

Table 6.8: (continued)

Desired property	How do we achieve it?	Effect	References
Low coating surface roughness	Use fine powder (<10 μm) feedstock	Higher degree of carburization	[6]
	Use nanograined feedstock	Lower coefficient of friction[a] at lower loads	[44]
Self-lubrication	Add fillers such as graphite, CNT, silver (Ag)	Improved tribological performance	[58–60]

[a]Friction, like wear, is a system property and is thus dependent on testing conditions and the counterface material.

The findings of this chapter can be summarized as follows:

- Of the various thermal spray techniques available, the combustion method (HVOF predominantly, followed by HVAF) has been popularly employed to spray apply CrC–NiCr coatings. PS and laser deposition techniques have also been used but to a much lesser extent. This chapter did not reveal any works involving production of CrC–NiCr coatings using the electric wire-arc method.
- The thermal spray route and process parameters adopted can result in varying microstructural characteristics which can subsequently impact tribological performance. Authors have also resorted to using popular wear models to characterize microstructure and analyze wear behavior.
- Both customized and standardized tribological tests have been carried out over the years by various researchers and have been reviewed thoroughly in this work. It is imperative that screening tests carried out on any coating system closely mirror the actual scenario.
- In terms of the scale at which wear occurs, sliding wear tests occur on a smaller length scale, and performance at this level has been found to be influenced by intralamellar properties.
- Wear tests conducted in abrasive or erosive modes occur at larger length scales and are as such predominantly influenced by interlamellar properties such as porosity and cohesion. Contrary to popular belief that an increase in hardness should correspond to an increase in wear resistance, this chapter has shown that this may hold true only in some cases and that too for tests carried out at higher length scales.
- Precoating activities such as selection of nanosized powder feedstock grains to application of engineering design methodologies and addition of transition layers have been found to enhance mechanical and tribological properties. In certain cases, various types of fillers such as nickel-cladded graphite, carbon nanotubes, silver and CaF_2/BaF_2 have been added to the CrC–NiCr matrix to render self-lubricating properties to the same.

- Postcoating activities such as heat treatment have led to improvements in erosion resistance as confirmed through various studies. Furthermore, laser posttreatments have helped in the production of dense microstructures that have, in certain cases, also led to enhanced fracture toughness and better resistance to crack propagation when subjected to abrasive wear environments.
- RCF studies carried out by some researchers have shown that the location of maximum stress amplitude is an important factor affecting the fatigue performance and failure mode of CrC–NiCr coatings. Furthermore, the type of spray technique adopted has been found to be an important factor affecting the amount of residual stresses and chance for crack propagation through the coating matrix. In this regard, the HVOF technique has been found to be better than the PS method for producing coatings with higher fatigue resistance. A low surface roughness has also been recommended for ensuring better fatigue life; however, more work remains to be carried out in this regard.

References

[1] Fauchais P, Vardelle A. Thermal sprayed coatings used against corrosion and corrosive wear. Adv Spray Appl 2012;3–39. DOI:10.5772/1921.
[2] Mcdonald AG. Analysis of thermal spraying in the industries of Western Canada. ITSC 2015: International Thermal Spray Conference; Long Beach, CA; 2015.
[3] El I. Numerical analysis of erosion of the rotor labyrinth seal in a geothermal turbine. Geothermics 2002;31:563–77. DOI:10.1016/S0375-6505(02)00014-7.
[4] Shchurova AV. Modeling of the turbine rotor journal restoration on horizontal balancing machines. Procedia Eng 2016;150:854–9. DOI:10.1016/j.proeng.2016.07.134.
[5] El Rayes MM, Abdo HS, Khalil KA. Erosion – corrosion of cermet coating. Int J Electrochem Sci. 2013;8:1117–37.
[6] Picas JA, Forn A, Matthäus G. HVOF coatings as an alternative to hard chrome for pistons and valves. Wear 2006;261:477–84. DOI:10.1016/j.wear.2005.12.005.
[7] Canan UH, Yuk-Chiu L. HVOF Coating case study for power plant process control ball valve application. J Therm Spray Technol 2013;22:1184–92.
[8] Staia MH, Suárez M, Chicot D, Lesage J, Iost A, Puchi-cabrera ES. Cr_2C_3–NiCr VPS thermal spray coatings as candidate for chromium replacement. Surf Coat Technol Elsevier B.V.; 2013;220:225–31. DOI:10.1016/j.surfcoat.2012.07.043.
[9] Gisario A, Puopolo M, Venettacci S, Veniali F. Improvement of thermally sprayed WC–Co/NiCr coatings by surface laser processing. Int J Refract Met Hard Mater Elsevier Ltd; 2015;52:123–30. DOI:10.1016/j.ijrmhm.2015.06.001.
[10] Singh H, Singh B, Prakash S. Mechanical and microstructural properties of HVOF sprayed WC–Co and Cr_3C_2 – NiCr coatings on the boiler tube steels using LPG as the fuel gas. J Mater Process Technol 2006;171:77–82. DOI:10.1016/j.jmatprotec.2005.06.058.
[11] Zhang XC, Xu BS, Xuan FZ, Tu ST, Wang HD, Wu YX. Rolling contact fatigue behavior of plasma-sprayed CrC–NiCr cermet coatings. Wear 2008;265:1875–83. DOI:10.1016/j.wear.2008.04.048.
[12] Sidhu HS, Sidhu BS, Prakash S. Wear characteristics of Cr_3C_2 – NiCr and WC – Co coatings deposited by LPG fueled HVOF. Tribol Int Elsevier; 2010;43:887–90. DOI:10.1016/j.triboint.2009.12.016.

[13] Janka L, Norpoth J, Eicher S, Ripoll MR, Vuoristo P. Improving the toughness of thermally sprayed Cr_3C_2 – NiCr hard metal coatings by laser post-treatment. Mater Des 2016;98:135–42.
[14] Matthews S, James B, Hyland M. The role of microstructure in the mechanism of high velocity erosion of Cr_3C_2-NiCr thermal spray coatings: part 1 – as-sprayed coatings. Surf Coat Technol Elsevier B.V.; 2009;203:1086–93. DOI:10.1016/j.surfcoat.2008.10.005.
[15] Vackel A, Dwivedi G, Sampath S. Structurally integrated, damage-tolerant, thermal spray coatings. J Miner Met Mater Soc 2015;67:1540–53. doi:10.1007/s11837-015-1400-1.
[16] Zang C, Wang Y, Zhang Y-D, Li J-H, Zeng H, Zhang D-Q. Microstructure and wear-resistant properties of NiCr – Cr_3C_2 coating with Ni45 transition layer produced by laser cladding. Rare Met. Nonferrous Metals Society of China; 2015;34:491–7. DOI:10.1007/s12598-015-0492-7.
[17] Fauchais PL, Heberlein JV, Boulos MI. Thermal spray fundamentals – from powder to part. New York, USA: Springer, 2014.
[18] Otsubo F, Era H, Kishitake K, Uchida T. Properties of Cr_3C_2-NiCr cermet coating sprayed by high power plasma and high velocity oxy-fuel processes. J Therm Spray Technol 2000;9:499–504. DOI:10.1007/BF02608553.
[19] Godwin G, Jaisingh SJ, Priyan MS. Tribological and corrosion behavior studies on Cr_3C_2 – NiCr NiCr powder coating by HVOF spray method- a review. J Mater Sci Surf Eng 2017;5:537–43.
[20] Kaur M, Singh H. A Survey of the Literature on the use of high velocity oxy-fuel spray technology for high temperature corrosion and erosion-corrosion resistant coatings. Anti-Corros Methods Mater 2008;55:86–96. DOI:10.1108/00035590810859467
[21] Shamim T, Xia C, Mohanty P. Modeling and analysis of combustion assisted thermal spray processes. Int J Therm Sci. 2007;46:755–67. DOI:10.1016/j.ijthermalsci.2006.10.005.
[22] Li M, Shi D, Christofides PD. Modeling and control of HVOF thermal spray processing of WC-Co coatings. Powder Technol 2005;156:177–94. DOI:10.1016/j.powtec.2005.04.011.
[23] Li M, Christofides PD. Computational study of particle in-flight behavior in the HVOF thermal spray process. Chem Eng Sci 2006;61:6540–52. DOI:10.1016/j.ces.2006.05.050.
[24] Ma XQ, Gandy DW, Frederick GJ. Innovation of ultrafine structured alloy coatings having superior mechanical properties and high temperature corrosion resistance. J Therm Spray Technol 2008;17:933–41. DOI:10.1007/s11666-008-9263-4.
[25] Kai TAO, Jie Z, Hua CUI. Fabrication of conventional and nanostructured NiCrC coatings via HVAF technique. Trans Nonferrous Met Soc China 2008;18:262–9.
[26] Oksa M, Turunen E, Suhonen T, Varis T, Hannula S-P. Optimization and characterization of high velocity oxy-fuel sprayed coatings: techniques, materials, and applications. Coatings 2011;1:17–52. DOI:10.3390/coatings1010017
[27] Zhang X. Several fundamental researches on structural integrity of plasma-sprayed coating-based systems. Weld World 2013;57:189–202. DOI:10.1007/s40194-012-0014-2.
[28] Vuoristo P, Tuominen J, Nurminen J. Laser coating and thermal spraying – process basics and coating properties. ITSC 2005: Thermal Spray connects: explore its surfacing potential. 2005,. 1270–7.
[29] Venkatesh L, Samajdar I, Tak M, Doherty RD, Gundakaram RC, Prasad KS, et al. Microstructure and phase evolution in laser clad chromium carbide-NiCrMoNb. Appl Surf Sci. Elsevier B.V.; 2015;357:2391–401. DOI:10.1016/j.apsusc.2015.09.260.
[30] Stanisic J, Kosikowski D, Mohanty PS. High-speed visualization and plume characterization of the hybrid spray process. ITSC 2006: Thermal Spray 2006: Science, Innovation, and Application. 2006, 1021–6. DOI:10.1361/105996306X147036.
[31] Matthews S, James B, Hyland M. The role of microstructure in the mechanism of high velocity erosion of Cr_3C_2 – NiCr thermal spray coatings: part 2 – heat treated coatings. Surf Coat Technol Elsevier B.V.; 2009;203:1094–100. DOI:10.1016/j.surfcoat.2008.10.013.

[32] Zhang XC, Xu BS, Tu ST, Xuan FZ, Wang HD, Wu YX. Fatigue resistance and failure mechanisms of plasma-sprayed CrC – NiCr cermet coatings in rolling contact. Int J Fatigue Elsevier Ltd; 2009;31:906–15. DOI:10.1016/j.ijfatigue.2008.10.006.
[33] Picas JA, Forn A, Igartua A, Mendoza G. Mechanical and tribological properties of high velocity oxy-fuel thermal sprayed nanocrystalline CrC-NiCr coatings. Surf Coatings Technol. 2003;174–175:1095–100. DOI:10.1016/S0257-8972(03)00393-1.
[34] Lih WC, Yang SH, Su CY, Huang SC, Hsu IC, Leu MS. Effects of process parameters on molten particle speed and surface temperature and the properties of HVOF CrC/NiCr coatings. Surf Coatings Technol 2000;133–4:54–60. DOI:10.1016/S0257-8972(00)00873-2.
[35] Mateos J, Cuetos JM, Vijande R, Fernández E. Tribological properties of plasma sprayed and laser remelted 75/25 Cr_3C_2/NiCr coatings. Tribol Int 2001;34:345–51. DOI:10.1016/S0301-679X(01)00023-8.
[36] Bolelli G, Berger LM, Börner T, Koivuluoto H, Lusvarghi L, Lyphout C, et al. Tribology of HVOF-and HVAF-sprayed WC-10Co4Cr hard metal coatings: a comparative assessment. Surf Coatings Technol Elsevier B.V.; 2015;265:125–44. DOI:10.1016/j.surfcoat.2015.01.048.
[37] Bolelli G, Berger L, Börner T, Koivuluoto H, Matikainen V, Lusvarghi L, et al. Sliding and abrasive wear behaviour of HVOF- and HVAF-sprayed Cr_3C_2 – NiCr hard metal coatings. Wear Elsevier; 2016;358–359:32–50. DOI:10.1016/j.wear.2016.03.034.
[38] Mruthunjaya M, Parashivamurthy KI. Microstructural characterization and hot erosion behavior of WC-12Co coated stainless steel using HVOF technique. Int J Mech Eng Technol 2016;7:53–62.
[39] Mruthunjaya M, Parashivamurthy KI, Devappa. Microstructural Characterization and Hot Erosion Behavior of CrC-NiCr Coated Steel Using HVOF Technique. Int J Mech Eng Technol. 2016;7:53–62.
[40] Wielage B, Pokhmurska H, Wank A, Riesel G, Stienhaeuser S. Influence of thermal spraying method on the properties of tungsten carbide coatings. Conference of Modern Wear And Corrosion Resistant Coatings Obtained by Thermal Spraying, Warsaw, Poland; November 20–21, 2003. Available at: https://www.gtv-mbh.com/_old/gtv-mbh-englisch/www.gtv-mbh.de/cms/upload/publikat/Wank/english/2003_06_eng.pdf
[41] Lima MM, Godoy C, Modenesi PJ, Avelar-Batista J., Davison A, Matthews A. Coating fracture toughness determined by Vickers indentation: an important parameter in cavitation erosion resistance of WC–Co thermally sprayed coatings. Surf Coatings Technol. 2004;177–178:489–96. DOI:10.1016/S0257-8972.
[42] López Báez I, Poblano Salas CA, Muñoz Saldaña J, Trápaga Martínez LG. Effects of the modification of processing parameters on mechanical properties of HVOF Cr_2C_3-25NiCr coatings. J Therm Spray Technol 2015;24:938–46. DOI:10.1007/s11666-015-0255-x.
[43] Matikainen V, Koivuluoto H, Milanti A, Vuoristo P. Advanced coatings by novel high-kinetic thermal spray processes. Materia 2015;73:46–50.
[44] Roy M, Pauschitz A, Polak R, Franek F. Comparative evaluation of ambient temperature friction behaviour of thermal sprayed Cr_3C_2– 25 (Ni20Cr) coatings with conventional and nano-crystalline grains. Tribol Int 2006;39:29–38. DOI:10.1016/j.triboint.2004.11.009.
[45] Wank A, Wielage B, Reisel G, Grund T, Friesen E. Performance of thermal spray coatings under dry abrasive wear conditions. 4th International Conference, THE Coatings, 2004. pp. 507–14.
[46] Wielage B, Wank A, Pokhmurska H, Friesen E, Grund T, Chemnitz D, et al. Correlation of microstructure with abrasion and oscillating wear resistance of thermal spray coatings. Thermal Spray 2005: Proceedings of the International Thermal Spray Conference, Basel, Switzerland, May, 2005:2–4.
[47] Nurbaş M, Ataba Durul EN. Abrasive wear behavior of different thermal spray coatings and hard chromium electroplating on A286 super alloy. Adv Mater Phys Chem 2012;2:68–70. DOI:10.4236/ampc.2012.24B019.

[48] Sosnowy P, Góral M, Kotowski S, Hanula G, Gwizdała J, Drzał J, et al. The influence of temperature on erosion resistance of carbide coatings deposited by APS method. Solid State Phenom 2015;227:251–4. DOI:10.4028/www.scientific.net/SSP.227.251.
[49] Fischer F, Dvorak M, Siegmann S. Development of ultra thin carbide coatings for wear and corrosion resistance. Thermal Spray 2001: New surfaces for a new millennium: Proceedings of the International Thermal Spray Conference, 2001. pp. 1131–5.
[50] Santa JF, Espitia LA, Blanco JA, Romo SA, Toro A. Slurry and cavitation erosion resistance of thermal spray coatings. Wear 2009;267:160–7. DOI:10.1016/j.wear.2009.01.018
[51] Sugiyama K, Nakahama S, Hattori S, Nakano K. Slurry wear and cavitation erosion of thermal-sprayed cermets. Wear 2005;258:768–75. DOI:10.1016/j.wear.2004.09.006.
[52] Matikainen V, Niemi K, Koivuluoto H, Vuoristo P. Abrasion, erosion and cavitation erosion wear properties of thermally sprayed alumina based coatings. Coatings 2014;4:18–36. DOI:10.3390/coatings4010018.
[53] Kim S-J, Lee S-J, Kim I-J, Kim S-K, Han M-S, Jang S-K. Cavitation and electrochemical characteristics of thermal spray coating with sealing material. Trans Nonferrous Met Soc China 2013;23:1002–10. DOI:10.1016/S1003-6326(13)62559-5.
[54] Kumar RK, Kamaraj M, Seetharamu S, Pramod T, Sampathkumaran P. Effect of spray particle velocity on cavitation erosion resistance characteristics of HVOF and HVAF processed 86WC-10Co4Cr hydro turbine coatings. J Therm Spray Technol Springer US; 2016;25:1217–30. DOI:10.1007/s11666-016-0427-3.
[55] Roy M, Pauschitz A, Bernardi J, Koch T, Franek F. Microstructure and mechanical properties of HVOF sprayed nanocrystalline Cr_3C_2-25(Ni20Cr) coating. J Therm Spray Technol 2006;15:372–81.
[56] Racek O. Wear resistant amorphous and nanocomposite coatings. Lawrence Livermore National Laboratory (LLNL) Livermore, California LLNL-JRNL-402714, 2008; 1–18. Available at: https://digital.library.unt.edu/ark:/67531/metadc831351/m1/1/
[57] Bobby S, Samad MA. Enhancement of tribological performance of epoxy bulk composites and composite coatings using micro/nano fillers: a review. Polym Adv Technol 2016;DOI:10.1002/pat.3961.
[58] Corte CD, Sliney HE. Composition optimization of self-lubricating chromium-carbide-based composite coatings for use to 760 °C. A S L E Trans Taylor & Francis; 1987;30:77–83. DOI:10.1080/05698198708981733.
[59] Sliney HE. The Use of silver in self-lubricating coatings for extreme temperatures. A S L E Trans Taylor & Francis; 1986;29:370–6. DOI:10.1080/05698198608981698
[60] Bartuli C, Valente T, Casadei F, Tului M. Advanced thermal spray coatings for tribological applications. J Mater Des Appl 2007;221:175–86. DOI:10.1243/14644207JMDA135.
[61] Singh V, Diaz R, Balani K, Agarwal A, Seal S. Chromium carbide-CNT nanocomposites with enhanced mechanical properties. Acta Mater 2009;57:335–44. DOI:10.1016/j.actamat.2008.09.023.
[62] Ksiazek M, Boron L, Radecka M, Richert M, Tchorz A. Mechanical and tribological properties of HVOF-sprayed (Cr3C2-NiCr+Ni) composite coating on ductile cast iron. J Mater Eng Perform 2016;25:3185–93. DOI:10.1007/s11665-016-2226-x.
[63] Picas JA, Punset M, Menargues S, Campillo M, Teresa Baile M, Forn A. The influence of heat treatment on tribological and mechanical properties of HVOF sprayed CrC-NiCr coatings. Int J Mater Form 2009;2:225–8. DOI:10.1007/s12289-009-0466-0.
[64] Rauf MM, Shahid M, Khan AN, Mehmood K. Laser cladding to improve oxidation behavior of air plasma-sprayed Ni-20Cr coating on stainless steel substrate. J Mater Eng Perform Springer US; 2015;24:3651–7. DOI:10.1007/s11665-015-1632-9

[65] Rauf MM, Shahid M, Durrani YA, Khan AN, Akhter AHR. Cladding of Ni–20Cr coatings by optimizing the CO_2 laser parameters. Arab J Sci Eng 2016;41: 2353–62. DOI:10.1007/s13369-015-1972-7.

[66] Rajendran R, S.S.K. B, Tamilselvi M, Chitra J, Bagade VU. Deformation and fracture behaviour of chromium carbide-nickel chromium wear resistant coating. 6th Asian Thermal Spray Conference, 2014. pp. 25–27.

[67] Zhang XC, Xuan FZ, Xu JS, Tu ST, Xu BS. Stress-dependent fatigue mechanisms of CrC-NiCr coatings in rolling contact. Fatigue Fract Eng Mater Struct 2011;34:438–47. DOI:10.1111/j.1460-2695.2011.01538.x

[68] Zhang X, Xuan F, Tu S, Xu B, Wu Y. Durability of plasma-sprayed Cr_3C_2-NiCr coatings under rolling contact conditions. Front Mech Eng 2011;6:118–35. DOI:10.1007/s11465-011-0127-0.

[69] Nieminen R, Vuoristo P, Niemi K, Mantyla T, Barbezat G. Rolling contact fatigue failure mechanisms in plasma and HVOF sprayed WC-Co coatings. Wear 1997;212:66–77.

[70] Varis T, Suhonen T, Calonius O, Čuban J, Pietola M. Optimization of HVOF Cr_3C_2-NiCr coating for increased fatigue performance. Surf Coat Technol 2016;305:123–31. DOI:10.1016/j.surfcoat.2016.08.012.

[71] Agrianidis P, Anthymidis KG, David C, Tsipas DN. Impact testing a capable method to investigate the fatigue resistance. Fract Nano Eng Mater Struct 2006;219–20. Available at: https://doi.org/10.1007/1-4020-4972-2_107

[72] Lima RS, Marple BR. Thermal spray coatings engineered from nanostructured ceramic agglomerated powders for structural, thermal barrier and biomedical applications: a review. J Therm Spray Technol 2007;16:40–63. DOI:10.1007/s11666-006-9010-7.

Index

https://doi.org/10.1515/9783110352986-007